```
TA   Blilie, Charles L.
165     The image as
.B57 information : a gen
     theory of sensors a
     sense-data / by Cha
```

THE IMAGE AS INFORMATION

A GENERAL THEORY OF SENSORS AND SENSE-DATA

THE IMAGE AS INFORMATION

A GENERAL THEORY OF SENSORS AND SENSE-DATA

Charles L. Blilie

NOVA SCIENCE PUBLICATION, INC
COMMACK, NY

Editorial Production: Susan Boriotti
Assistant Vice President/Art Director: Maria Ester Hawrys
Office Manager: Annette Hellinger
Graphics: Frank Grucci
Acquisitions Editor: Tatiana Shohov
Book Production: Ludmila Kwartiroff, Christine Mathosian, Joanne Metal and Tammy Sauter
Circulation: Iyatunde Abdullah, Sharon Britton, and Cathy DeGregory

Library of Congress Cataloging-in-Publication Data available upon request

ISBN 1-56072-458-7

Copyright © 1997 by Nova Science Publishers, Inc.
6080 Jericho Turnpike, Suite 207
Commack, New York 11725
Tele. 516-499-3103   Fax 516-499-3146
E-Mail: Novascience@earthlink.net

All rights reserved. No part of this book may be reproduced, stored in a retrieval system or transmitted in any form or by any means: electronic, electrostatic, magnetic, tape, mechanmical, photocopying, recording or otherwise without permission from the publishers.

The authors and publisher haven taken care in preparation of this book, but make no expressed or implied warranty of any kind and assume no responsibility for any errors or omissions. No liability is assumed for incidental or consequential damages in connection with or arising out of information contained in this book.

This publication is designed to provide accurate and authoritative information with regard to the subject matter covered herein. It is sold with the clear understanding that the publisher is not engaged in rendering legal or any other professional services. If legal or any other expert assistance is required, the services of a competent person should be sought. FROM A DECLARATION OF PARTICIPANTS JOINTLY ADOPTED BY A COMMITTEE OF THE AMERICAN BAR ASSOCIATION AND A COMMITTEE OF PUBLISHERS.

Printed in the United States of America

Contents

LIST OF FIGURES AND TABLES ix

INTRODUCTION xi

PART I THE DESCRIPTION AND CAPABILITIES OF SENSORS

SENSORS: A GENERAL OVERVIEW 3
1.1 The Process of Sensation 3
1.2 Sensing versus Perceiving 6
1.3 Kinds of Sensors and Information 9
1.4 Modes of Physical Interaction 11
1.5 Sense Organs 17
1.6 Artificial Sensors 24

PROTOSENSORS AND SIMPLE SENSORS 27
2.1 The Meaning of Sensory Transduction 27
2.2 The Protosensor and its Characteristics 29
2.3 Protosensors and Knowledge 31
2.4 The Simple Sensor 37

COMPLEX SENSORS 41
3.1 Kinds of Multielement Sensor 41
3.2 Complex Sensor Synthesis 43
3.3 The Image as an Array of Intensities and Qualities 51

3.4 EXTENSION AND INTENSION OF COMPLEX SENSORS	66
MULTISENSOR FUSION	**69**
4.1 GENERAL ASPECTS OF MULTISENSOR FUSION	69
4.2 GROUNDS FOR MULTISENSOR DATA FUSION	71
4.3 SPATIO-TEMPORAL FUSION	72
4.4 INTRINSIC FUSION	74
4.5 POSITIONAL FUSION	75
4.6 BIOLOGICAL SENSORY INTEGRATION	76
4.7 THE MULTIMODE SENSOR IMAGE	78
4.8 THE ARGUS SENSOR	79
TEMPORAL CONSIDERATIONS AND VARIABLE SENSORS	**81**
5.1 DYNAMIC ASPECTS OF SENSATION	81
5.2 DYNAMIC RESPONSE OF SENSORS	82
5.3 ANALYSIS OF THE DATA STREAM	83
5.4 TRADING TEMPORAL FOR NON-TEMPORAL	85
5.5 VARIABLE SENSORS	87
5.6 UNITY AND DIVERSITY OF THE SERIES OF FRAMES	88
ACTIVE SENSORS	**91**
6.1 ACTIVE SENSING IN GENERAL	91
6.2 ACTUATORS	93
6.3 COMBINATION OF ACTUATORS AND SENSORS	94
6.4 ACTIVE SENSORS AND KNOWLEDGE	95
6.5 THE ACTIVE ARGUS SENSOR	98
PART II THE MEANING AND VALIDITY OF SENSORY INFORMATION	
THE SENSORY OBJECT	**103**
7.1 GENERAL CONSIDERATIONS	103
7.2 REPRESENTING THE SENSORY OBJECT THROUGH PROTO-OBJECTS	104
7.3 CORRESPONDENCE OF SENSORY AND PHYSICAL OBJECTS	107
7.4 INTELLIGENT EMITTERS	110
PERSPECTIVE AND ERRORS IN SENSATION	**113**
8.1 FROM DISTAL TO PROXIMAL STIMULUS	113

8.2 Projection, Perspective and Stimulus Dynamics	115
8.3 Errors from the Medium	122
8.4 Errors from Sensors	125
8.5 Strategies for Dealing with Errors	130
8.6 Summary: An Information Theory Model of the Relation of Sensor and Object	132

QUALITIES AND THEIR RELATION TO OBJECTS — 135

9.1 The Problem of Qualities	135
9.2 The Aristotelian Theory of Sense Qualities	137
9.3 Modern Thought on Qualities	140
9.4 The Objectivity of Qualities in General Sensor Theory	143
9.5 Qualities and Perception	154
9.6 Intentionality and the Mutual Conformation of Sensor and Object	160

IMAGE PROCESSING AND PATTERN RECOGNITION — 167

10.1 Preliminary	167
10.2 Operations on a Single Image	169
10.3 Image Segmentation and Analysis	175
10.4 Multi-Image Operations -- the Comparison and Contrast of Images	179
10.5 Pattern Recognition and Object Identification	182
10.6 Abstraction and Concept Formation	185

SENSIBILITY IN GENERAL — 191

11.1 The Problem of Sensibility in General	191
11.2 Sensibility and Systemic Relations	192
11.3 Applications	196

GLOSSARY OF TERMS — 199

NOTES AND REFERENCES — 213

INDEX — 221

LIST OF FIGURES AND TABLES

Figure 1-1	The Sensory Process – Idealized	4
Figure 1-2	The Sensory Process – Realistic	7
Figure 1-3	The Sensory Scene from the Subject's Viewpoint	8
Figure 3-1	Spatial Arrangements of Sensor Elements	44
Figure 3-2	Response of Human Photoreceptors as a Function of Wavelength	50
Figure 3-3	Infrared Detector Response Function	54
Figure 3-4	Translation of Light into Color Hue and Saturation	59
Figure 3-5	Polarization States and Their Translation	61
Figure 5-1	Analysis of the Data Stream into Frames	85
Figure 6-1	Integrated and Separated Active Sensors -- a Schematic	97
Figure 8-1	Varieties of Sensor Errors	126
Figure 8-2	The Relation of Sensor and Object as a Communications System	133
Figure 10-1	Visual Receptive Fields and the $\nabla^2 G$ Operator	171
Figure 10-2	Optical Illusions Resulting from Edge Enhancement	173
Figure 10-3	Examples of Images and their Fourier Transforms	175
Figure 10-4	Cluster Analysis and the Formation of Reference Patterns	189
Table 1-1.	Physical Force Fields	13
Table 1-2.	Physical Propagating Objects	14

Table 1.3.	Propagating Waves.	15
Table 1-4.	The Human Senses	19
Table 1-5.	Artificial Sensors of Force Fields	25
Table 1-6.	Artificial Sensors for Propagating Objects.	26
Table 1-7.	Artificial Sensors for Propagating Waves.	26
Table 3-1.	Intensity and Qualities: A Summary	53
Table 10-1.	Comparison of Sensory Information with Perceived Information	188

INTRODUCTION

Sensors are changing our world as profoundly as computers have, yet they have not received nearly the same attention. From a practical beginning with the lowly thermostat that controls your furnace or refrigerator, sensors are now used in televisions to receive remote control commands, in the motion detectors of alarm systems, and even to operate faucets. The 20th Century has seen a virtual explosion of sensor technology. The commercial and medical applications of artificial sensor technology are too numerous to list; they can only be categorized. Since automation requires immediate information regarding objects, the growth of sensor technology has been essential for robotics. Sensors have been a key area of research for the military. They are the eyes and ears of armies, navies, and air forces, and they have sometimes provided a decisive strategic edge, as radar did for the Allies in the Second World War. Military applications have driven tremendous advances in the technologies of radar, sonar, and imaging infrared sensors, many of which now find civilian employment. The scientific uses of sensors are equally numerous; in many ways, artificial sensors and experimental science have grown in unison. Sensors -- like telescopes and microscopes -- allow the scientist to peer into regions invisible to the unaided human senses.

Modern science has also made vast strides in unlocking the secrets of the physiology and psychology of sensation. It has elucidated the processes of sensory transduction at the microscopic level, the encoding and communication of sensations in the nervous system, and the representation of sensory information in the brain. It has discovered animal senses unlike those of human beings, such as echolocation in bats and infrared vision in snakes. And scientific research has also brought out remarkable parallels between biological sense organs and artificial sensors.

Yet the problem of sensation and sense perception is not at all new. It grew up within epistemology, the branch of philosophy that deals with knowledge. Speculation on the nature and validity of sensation is probably as old as philosophy itself. And for good reason: sensation is necessarily our source of knowledge about the world and the multitude of particular things it contains. Views on the epistemological status of sense perception are dividing lines between entirely different philosophic schools. Despite this importance, philosophic works on sensation have drawn upon but little of recent scientific study regarding sensors and sensory processes and have tended instead to rehash old problems from the point of view of the unaided human senses.

What is needed is a *general* theory of sensors that ties all these fields together in a unified approach. The purpose of this book is to outline such a theory, its methods, and some of its results. A general theory of sensors and sensory information draws upon several specialized fields, each of which has a different perspective on the problem. It would be impossible for this work to exhaustively cover every topic raised in a consideration of sensors. Rather, its intent is to establish a unified approach and explanatory framework, and to stimulate further research.

The specialized scientific fields -- such as sensor engineering, the physiology and psychology of perception, and artificial intelligence -- have all advanced more or less independently of each other. General sensor theory here provides a solid, unified foundation, as many of these fields never rise to theoretical reflection. They have aimed at practical results, and texts in these fields, especially for artificial sensors, are often at the nuts-and-bolts practical level. At the same time, general sensor theory is necessary to epistemology to prod it beyond human sensation to a truly general perspective. Ultimately, it hopes to understand the possible limits and nature of sensory knowledge. General sensor theory provides a hub which unites all these fields, and makes fruitful communication between them possible. Given the tremendous strides made in sensor technology and sensory physiology in the past few decades, a unified understanding is essential for the promises of future developments.

Given the two inclinations present in the study of sensors -- the practical and the purely theoretical -- this book has a twofold purpose, which is reflected in its organization. First, it presents a completely general theory of sensors and sensation, with an emphasis on their structure and description. It shows how to describe any sensor or image in terms of digital information. It then applies this theoretical approach to some longstanding problems in sense perception.

Sensor theory has many uses and benefits. It provides a means for precise characterization of the information content of images and a way of describing sense-qualities. It shows how information theory and models of communications can be applied to the sensory process. It will assist in the important and difficult problem of multi-sensor data fusion. It will suggest possible new kinds of sensors and show the limits and meaning of information from such sensors. Sensor theory could also assist in the choice of the most advantageous scheme for the representation of sensory information in a computer system.

In physiology and psychology, sensor theory will, as in the study of artificial sensors, provide a framework for the understanding of sensory information, as well as the comparison of human and animal senses. It might also be of assistance in the development of artificial sensors for use in human beings. It can provide an approach to the translation of other sensory information (infrared vision, for example) into optimal displays for human sensing. This could be of use in difficult ergonomics problems, such as the display of information in fighter aircraft.

For the epistemologist, sensor theory provides a new basis for the discussion of sensation and perception. Primarily, it allows epistemology to readily draw upon developments in other fields, and it expands discussion of sensation beyond the purely human. General sensor theory also introduces some important new ideas for philosophic reflection, such as the theory of the protosensor, the problem of multimode sensor fusion, active sensing, the parsing of a data stream into discrete frames, stimulus dynamics, image processing, and the Argus sensor.

Artificial intelligence and robotics have already expended great efforts in the areas of computer recognition of patterns, formation of classes and concepts, and even into whether computers will be able to think. Yet they have given surprisingly little attention to sensors and the meaning of sensory information. General sensor theory remedies this defect. The area known as computer vision should especially benefit from general sensor theory. General sensor theory may also find application to the artificial generation of images, both in computer graphics and in virtual realities. As we will see, the same approach can used for generating image information as receiving it. For virtual reality designers of the future, general sensor theory provides a means of estimating just how much information must be displayed in order to make it appear real to a human observer.

This book is *not* an engineering manual for the design of artificial sensors, nor is it a discussion of the details of human sense organs or sensation. Many books exist on these topics, and they will be referred to where necessary in the present work. The aim here is to draw upon these detailed studies and find a theory which encompasses them. Neither is this book a purely philosophical view of sense perception. It is the common mistake of epistemology to jump immediately to a theoretical study of sense perception without having clearly distinguished perception from the sensory processes on which it is based. Philosophic readers thus should not lose patience with the scientific examples given, but, on the contrary, view them as material for reflection. Sensor theory is a necessary preliminary to the study of perception. Scientists, engineers, and computer programmers reading this book may at times become impatient with what seems to them to be pointless epistemological hairsplitting. Confident in their own common sense, down-to-earth view of reality, they may wish to skip these sections in a first reading. Yet they should realize that such theoretical argumentation is necessary if sensor theory is to be well-founded and truly general. Properly viewed, general sensor theory is to the sensor analyst what physics and mathematics are to mechanical or electrical engineer -- an overarching framework through which innumerable practical applications can be understood.

This book's approach to general sensor theory, to repeat, divides naturally into two parts. Part I (Chapters 1 through 6) deals with the description and composition of sensors and the characterization of the information produced by sensors, whether they be artificial sensors or the sense organs of living beings. First of all, the nature of the sensory problem is discussed, the distinction is drawn between sensing and perceiving, and the possible kinds of sensors and sense organs are enumerated. The aim here is to define what a sensor is and what methods are used in sensing. Next, the book introduces the theory of the protosensor, and shows how all sensors may be represented as a combination of infinitesimal protosensors. This is key to understanding the meaning of sensory images and the validity of such images. Complex sensors are then discussed, and this treatment is extended to multimode sensor fusion, culminating in the Argus sensor. Then the book briefly considers active sensing and the dynamic aspects of sensation.

Part II (Chapters 7 through 11) deals with the theoretical and epistemological implications of sensor theory. These include the nature of the sensory object, projections and errors, an information theory model of the sensory process, the relation of qualities to the object, and image analysis. The problem of sensibility in general is touched upon and the similarities between general sensor theory and systems theory are explored. Finally, there is a glossary of technical terms used in this book, as several new terms are introduced.

PART I

THE DESCRIPTION AND CAPABILITIES OF SENSORS

CHAPTER ONE

SENSORS: A GENERAL OVERVIEW

1.1 THE PROCESS OF SENSATION

Sensors are best defined from their context: that is, from the central role they play in the sensation. *Sensation* is a process in which an object, through the intermediaries of a stimulus and a sensor, acts upon an information system. Figure 1-1 illustrates this in an idealized way. The term "information system" is taken with the broadest meaning. The human mind, the nervous system of a bee, and a radar tracking computer are all information systems for the purposes of sensor theory[1]. An information system contains *data representations* -- a set of numbers or symbols -- which it processes, combines, differentiates, and stores. Sensation is thus a particular kind of relation between a knower and an external object, which allows the object to evoke a data representation in the information system. For sensation to be possible, all the elements of the process -- object, stimulus, sensor, information system -- must be present. *Sensory information* is a data representation resulting from sensation. And let us call the capability to sense objects (the possibility of sensing) *sensibility*.

The crucial role of the sensor in the sensory process is obvious. It is the information system's window to the outside world. The sensor is the interface between the world of things and the information system with its data representations. Thus, a *sensor* translates a stimulus into some data representation

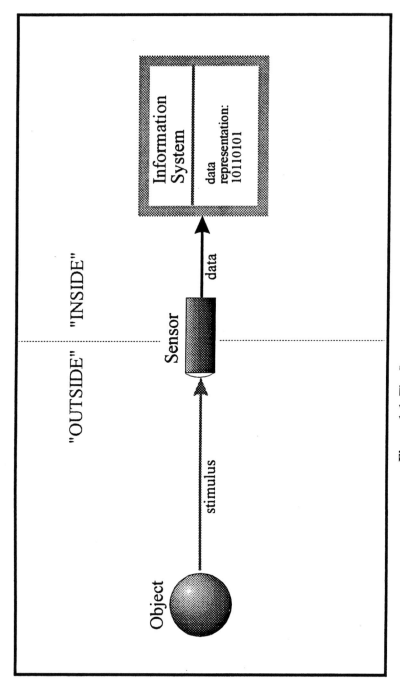

Figure 1-1. The Sensory Process -- Idealized.

within the information system. Put differently, a sensor converts physical energy into information[2,3]. (The distinction of data and information is considered below in Section 1.3.) A physical stimulus always involves a transfer of energy and causes a change in the sensor. The energy may be of any form: light, sound waves, chemical substances, radio waves, and so on. The information output of a sensor can be of many forms: an electrical current of a certain intensity, a binary number, or a series of nerve impulses. Such levels or numbers will have a specific meaning to the information system. The sensor output is thus a code or symbol, indicating its response to something in the physical world. The information symbolizes the type of stimulus.

This view of sensors makes some provisional assumptions regarding the nature of the sensory world[4]. For present purposes, it assumes with common sense that the sensory world and the multiplicity of objects it contains really exist: that they are not illusions or products of the mind. It further assumes that the sensory world is what physics and chemistry say it is: that it is composed of atoms, molecules, radiation, and so on. Thus, sensory stimuli are physical stimuli; that is, they are physical interactions.

A *sense organ* is a sensor possessed by a living being. Sense organs differ from artificial sensors only in that they are composed of cells and form an integral part of an organism. It may be disputed if plants have sensation (phototropism might be considered sensation of a primitive sort). It is certain, however, that all animals have one or more sense organs. Incidentally, this shows us the natural purpose of sensation. To respond to and act on particular things in the world requires information about them, and this particular information can only be gotten by sensors. Sensation is thus a precondition of action.

The fact that sensors translate energy into information distinguishes them from *transducers*. All sensors are transducers, but not the reverse. The energy output of a sensor is only a means to transfer information. Transducers, in contrast, are intended to transform one kind of energy into another. They do nothing with the information contained as such. A microphone is a transducer, for example, but could only be considered a sensor if its output had meaning to some information system. A radio is a transducer, but not a sensor, because it merely shifts the carrier of the information from radio waves to sound waves. It does nothing in the way of encoding or extraction of the information. A jackhammer and an electric generator are also transducers, but have no similarity to sensors. Neither are telescopes and microscopes sensors: they simply change our field of view and resolution thereof. *Instruments* are sensors if they actually receive information from the world. Instruments of their nature yield measurements, which are quantitative information. However, not all instruments are sensors, as they do not all convert energy into information. A meter stick, for example, is not a sensor.

Sensor theory is concerned with the origination, validity, representation, and meaning of sensory information. More to the point, it tries to understand how (or if) sensory information can be valid under conditions in the real world. The actual sensory process, illustrated in Figure 1-2, is far more complex than the idealized case previously described. The object will have a spectrum of physical interactions. These interactions mean the object reflects, absorbs, or emits energy: it is a bundle of potential stimuli (the *distal stimulus*)[5]. These physical interactions propagate in space through a medium, which may contain other objects that obscure or distort the stimuli from the object. Plus, they can add stimuli of their own. All the stimuli are projected (the *proximal stimulus*) on to the sensor. The sensor translates the proximal stimulus into a data representation. Sensor theory seeks to determine what relation and what similarity these data representations -- the sensory information -- have to the actual properties of the object.

1.2   Sensing versus Perceiving

Let us rotate our model of sensation around to the viewpoint of the subject. Forget for a moment everything that was said regarding sensors and stimuli. Putting aside all presuppositions and theories, what is the sensory scene as perceived? If we open our senses to the world , what do we perceive? It is something rather different than an abstract description of stimuli and data representations. We perceive spatial regions "out there", with various qualities (colors, shapes, sounds, textures, etc.) that come and go or stay unmoved (Figure 1-3). Sensor theory must ultimately connect these qualities and regions with the natures of objects.

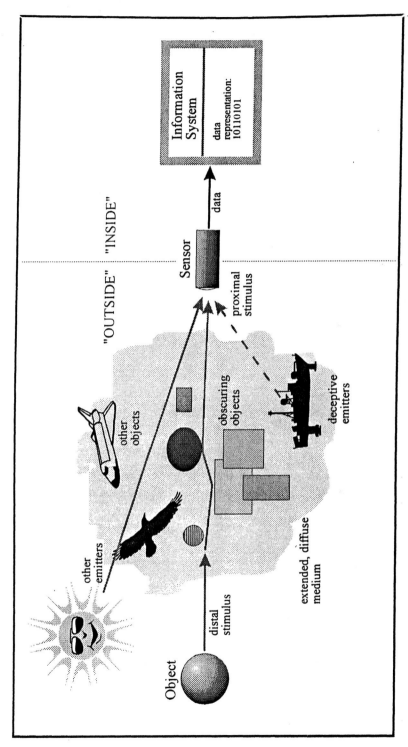

Figure 1-2. The Sensory Process -- Realistic.

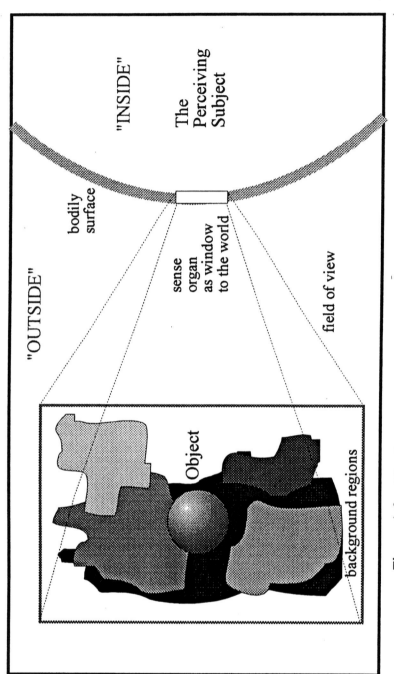

Figure 1-3. The Sensory Scene from the Subject's Viewpoint.

Sensor theory alone cannot explain why qualities are perceived as they are. It can tell us, for example, what data representation corresponds to red, what frequencies of light impinging on the eye lead to that representation, and what chemical substances reflect or emit those light frequencies. It cannot, however, tell us *why* red is red. Redness is the representation *as it is perceived*. It must be perceived to be understood.

This means sensor theory must draw a distinction between sensation and perception. The terms sensation and perception are often used interchangeably. Indeed, some philosophers and psychologists assert they are synonymous[6]. However, there are some salient points which differentiate them. Perception, unlike sensation, implies the participation of a knower, a conscious subject. It thus involves theories of mind, and properly belongs to psychology and epistemology. Sensation, in contrast, is something purely "mechanical" and unconscious. It is an information processing task that makes no assumptions regarding the nature of mind. We may dispute, for example, if artificial perception is possible, but there is no doubt that artificial sensation is possible. When we perceive the world, sensation takes place before perception. Perception is always based on it, rather than the reverse. Sensation is always of what is happening now and is stimulating us, while perception can be of memories, thoughts, and so on. Sensation merely collects information, while perception helps decide what to do with it. Finally, perception is always a covert process -- its workings cannot be directly observed, but must be inferred[7]. The details of sensation, on the other hand, can often be openly traced. All in all, sensation is something more primitive and more deterministic than perception, culminating in the formation of a data representation or image. It is to this domain that sensor theory limits itself.

1.3   KINDS OF SENSORS AND INFORMATION

There are important distinctions between kinds of sensors and the information they yield. These distinctions will be dealt with at length in later chapters; here, they are only introduced and defined.

One basic division is between *simple* and *complex* sensors. Simple sensors are the building blocks from which the complex sensors are constructed. They are monolithic and self-contained. We generally refer to simple sensors as *detectors*, *receptors*, or *sensor elements*. The rods and cones in the retina, the hair cells of the inner ear, and taste buds are all simple sensors. A single photodetector is an simple artificial sensor. A very important kind of simple sensor is the *protosensor*. Protosensors are idealized simple sensors that detect a single kind of energy; their importance lies in the fact that all sensors may be represented as a combination of infinitesimal protosensors. Chapter 2 explores the theory of the protosensor. Complex sensors, to repeat, are combinations of simple sensors. Human sense organs are complex sensors, each of which may contain millions of receptors. Complex sensors themselves fall into two classes: *multielement* and *multimode*. Multielement complex sensors combine a number of sensor elements of the same kind. Each of the human sense organs is a multielement sensor. A multimode sensor combines two or more sensors of different kinds. A multimode sensor may also be referred to as a *sensor suite* or *multisensor fusion*. The five human external senses in combination constitute a multimode sensor suite. An example of artificial multisensor fusion is the combination of an infrared imaging sensor with a radar in an aircraft. The problem of multisensor data fusion is quite difficult, and has been studied extensively. Chapters 3 and 4 examine complex sensors.

Sensor theory also distinguishes between *passive* and *active* sensors. All sensors require some kind of passivity, which is to say they can be modified by stimuli. They are receptive, in other words. All sensation involves some change in the sensor -- it must receive or "capture" the physical stimulus in order to convert its information. A *passive sensor* is *purely* receptive: human vision and hearing are examples. Passive sensors collect information from their environs without acting on it; they simply observe. Since observation is the primordial and fundamental role of sensation, the vast majority of actual sense organs and artificial sensors are passive. In contrast, an *active sensor* can act upon or illuminate its object[8]. The classic example of an active sensor is radar. The radar sends out a radio beam, which is reflected by objects back to the radar receiver. The only human active sense is touch, especially in the hands. Touch is active because we intentionally reach out to touch things to learn more about them. Chapter 6 deals with the problem of active sensation at length.

Active sensors must have a way of acting on the object. This means there must be a device that does the opposite of a sensor. Instead of converting energy into information, it must convert information into energy. Such a device or organ is an *actuator*. An actuator acts as interface between an information system and the outside world. The theory of actuators in a general sense belongs to robotics and control theory and is thus beyond the scope of this book. The actuators of interest in sensation are those capable of bringing us new information about objects.

Finally, there is a division between *static* and *variable* sensors. The parameters of a static sensor remain constant, while those of a variable sensor can be adjusted to respond to changes in the outside world. A marvelous example of a variable sensor is the human eye. The iris changes the amount of light admitted to the eye, while the retina can adapt itself to very low light conditions in night vision. The eyes move in their sockets to change their viewpoint, while the lens flexes in order to change its depth of view. Variable sensors are considered in Chapter 5.

The purpose of any sensor, to repeat, is to bring us information about the world. The type of information yielded not surprisingly depends on the sensor type, the most important division being that between simple and complex sensors. The output of a simple sensor will be referred to as *data, information*, or *signal*. The output of a complex sensor may additionally be called an *image*. Both images and data constitute *representations* within the information system. The distinction between data and information has been drawn variously. In the present work, *data* will be taken to be a collection of facts, represented by symbols or numbers, regardless of whether they are of interest or not. (In science, data has the implication of being something quantitative; that is not the case here.) Data of actual interest to a system is *information*[9]. Information is meaningful data; sensory information is thus data that makes knowledge of sensible objects possible.

1.4   MODES OF PHYSICAL INTERACTION

Sensors are also classified according to *mode*. Each kind of physical interaction defines a possible mode of sensation. The five human external senses, for example, constitute five sensory modes: the eyes sense light, the ears sense compression waves in air or other fluids, smell and taste sense particular chemical substances, and touch senses mechanical contact. Each sensory mode, then, corresponds to a kind of physical interaction. The interaction is the *carrier* of information. (Even if we consider information to be in itself immaterial, it must be imposed on a physical vehicle if we are to sense it.) All sensing involves energy transfer from the object to the sensor. Both object and sensor must undergo some kind of change, and both must have a physical manifestation, in order for sensation to be possible.

Since the sum-total of the kinds of physical interactions defines all the possible modes of sensation, to classify these interactions is to also classify all possible sensors. There are only a few physical means of interaction, and these must be the carriers of the information. Physicists believe that all interactions can ultimately be reduced to the four basic forces -- electromagnetic, gravitational, weak, and strong -- and waves thereof (e.g., light is an electromagnetic wave). No physical information can be transferred faster than light speed. Sensor theory must also consider macroscopic (or secondary) interactions, such as waves in media (e.g., sound waves) and chemical or biological interactions. Each interaction has a specific kind of physical manifestation -- an electric field extending spherically in three dimensions from an electric charge, for example. The kinds of manifestation naturally divide physical interactions into three basic classes: *force fields*, *propagating objects*, and *propagating waves*.

Each interaction has a number of possible states it can assume. For example, electromagnetic radiation is specified by a frequency and a polarization, as well as by a direction of propagation. These will be called the *parameters* of the interaction. The parameters are what a sensor or instrument can actually measure.

1.4.1 FORCE FIELDS

A force field of its nature is extended in space and acts on objects that can respond to it. The best example is an electric field. On a macroscopic level, pressure forces also fall into this class. Detection of a force field generally requires the field to perform work on some object: the deflection of a compass needle, for example. Table 1-1 summarizes all known force fields. *Primary* force fields are direct manifestations of the fundamental forces of gravity, electromagnetism, and so on. They can occur in a vacuum. *Secondary* forces, such as pressure, depend on a medium like air or water. This distinction of primary and secondary also pertains to propagating objects and waves.

Table 1-1. Physical Force Fields.

class of force	physical manifestations	parameters
primary forces: macroscopic	electric fields magnetic fields gravity	magnitude, direction magnitude, direction magnitude, direction
primary forces: microscopic	weak force strong force	various
secondary forces: macroscopic	fluid pressure solid pressure or force (stress, strain, etc.)	magnitude magnitude, direction (tensor components)
secondary forces: microscopic	interatomic and intermolecular forces	various

The measurable parameters can vary in time, but such variability will usually give rise to propagating waves (see 1.4.3), especially if the variation is oscillatory. Of the primary forces, the ones of interest to the sensor designer are the macroscopic forces of electromagnetism and gravity. Human beings cannot directly sense electromagnetic forces, but our sense of balance does tell us our orientation with respect to gravity. The microscopic primary forces are of little relevance here, as one can detect them only on the subatomic level. The microscopic secondary forces manifest themselves in chemical reactions, which will be considered in the next section. The macroscopic secondary forces are all varieties of pressure. They are the basis of forces of contact between everyday objects and thus the sense of touch.

1.4.2 Propagating Objects

Interaction by propagating object involves an energy transfer by the motion of something from one place to another. Table 1-2 summarizes the varieties of propagating objects.

In other words, for propagating objects, there is always a dynamic in the interaction, unlike that for a static force field. (In chemical reactions, the energy transfer is in the reaction itself.) Measurement of propagating objects is invariably the measurement of a flux or rate. Of the subatomic objects, the most important by far is electric current; the others are of interest mainly to the physicist. Of the atomic and molecular transfers, the only one sensed directly by human beings is thermal flux, which we perceive as heat and cold. Macroscopic propagating objects are the area of interest of classical mechanics.

Table 1-2. Physical Propagating Objects.

class of object	physical manifestation	parameters
subatomic	electric current (electrons)	magnitude, direction
	fluxes of other elementary particles	number flux, energy, direction
	flux of nuclei (e.g., decay, cosmic rays)	number flux, energy, direction
atomic and molecular	atomic beams	number flux, energy, direction
	thermal flux (conduction or convection)	energy flux, direction
	chemical diffusion	rate, direction
	chemical reaction or absorption	rate
biological	transfer of cells or cell components	rate, direction
macroscopic	fluid flow	rate, direction
	solid contact -- friction or motion	momentum, angular momentum, acceleration

1.4.3 PROPAGATING WAVES

Propagating waves result from the oscillatory motion of objects or fields. These are of immense importance to sensors -- most sensors detect waves. Propagating waves bring information regarding distant objects in a more reliable fashion than either force fields or propagating waves. They are thus invariably the source of spatial information regarding far away objects: their locations, motion, shapes, and sizes. Table 1-3 summarizes the known varieties of propagating waves.

Table 1.3. Propagating Waves.

class of wave	physical manifestation	parameters
electromagnetic waves	radio waves (RF) microwaves infrared light (IR) visible light ultraviolet light (UV) x-rays gamma rays	intensity, frequency, polarization, direction, coherence
other primary waves	gravity waves (conjectured) vectors of weak and strong forces	various
secondary waves (waves in physical media)	sound waves in air in other fluids in solids microscopic (phonons)	intensity, frequency, direction
	surface waves	intensity, frequency, direction

Electromagnetic waves are probably the most important to sensors, since nearly everything in the universe emits, absorbs, and reflects electromagnetic radiation. They are thus an excellent way of discovering the properties of distant objects. Our most sophisticated sense -- vision -- is a detector of electromagnetic waves. Radio waves are, of course, what make radar possible. Radio frequency waves and microwaves can be generated by varying macroscopic electric currents, while infrared, visible, and ultraviolet light are typically produced by energy transitions within atoms and molecules. X-rays come from transitions of the innermost electrons in atoms, while gamma rays are produced by energy transitions within the atomic nucleus. Electromagnetic waves always travel at the speed of light (and along a straight line) in a vacuum. The other primary waves are of theoretical interest only. The vectors of the weak and strong forces travel only for subatomic distances, while gravity waves -- posited by the general theory of relativity -- still have not been directly observed or generated.

Secondary waves are of great importance to human beings, as we communicate verbally through sound. Sound waves are compression waves in various physical media, and their speed depends on the medium in question, as well as the density of the medium.

1.4.4 STIMULUS PARAMETERS AND THEIR MEASUREMENT

Within each mode of physical interaction, and hence each mode of possible sensation, there are the various parameters that are detected and measured. Light has an intensity, a frequency, a polarization, and a direction. These parameters define all possible *states* of the carrier, and allow us to draw distinctions between different carriers of the same mode. Visible light is composed of one or several frequencies of electromagnetic waves, for example. The states of the interaction define a set of sub-modalities within each mode, and these sub-modalities are detected or perceived as *qualities*. To return to example of light, the sub-modalities or qualities of light are perceived as colors[10]. Corresponding qualities in hearing are different pitches or tones. The meaning of qualities is dealt with at length in later chapters.

A sensor need not distinguish all the states of the carrier; it can ignore them or blend them together. Our eyes are not sensitive to the specific polarizations of light, for example, and even in the area of frequency, the eye sees only three primary colors and uses them to represent the others (we see multiple frequencies of light as a single color).

Any of the parameters of a physical stimulus can be measured, and these can be patterned or structured in space and time. Often, the primary purpose of a sensor is to detect precisely these spatial or temporal patterns and to ignore the details of the physical carrier itself. This determines how the sensor receives incoming stimuli. There is first a choice between an instantaneous measurement or a time-integrated one. In a practical sense, no sensor can respond instantaneously, so there will always be some time integration involved. The instantaneous values (e.g., flux) can still be estimated from a time series of integrated ones (e.g., fluence). Conversely, the sensor may use a long time integration to settle to an equilibrium value, which is then measured. Thermometers, for example, are often used in this way. A similar problem arises with regard to space. The sensor can measure properties at a single point or an array of points. Or, to the contrary, it can measure a flux through a surface.

One parameter present in all sensation is the *intensity* (magnitude or *quantity*) of the interaction. This merely expresses the fact that there must be an energy transfer in order for sensing to occur. Intensity will always have units of energy, energy flux, number, or number flux, and are non-negative. The qualities (and the mode itself) are specifiers of intensity, and the output of a sensor is an expression or representation of *intensity of a certain kind*. All actual sensors have an intensity threshold, an intensity value below which the sensor will not respond. Actual sensors also have a maximum intensity -- a saturation or damage point -- above which the sensor cannot measure.

1.5 SENSE ORGANS

To repeat, sensation is found in all animals and is necessary for motion and action. And, as we might expect, animals have many different varieties of sense organs, suited to their needs. Despite the variety, most sense organs fall into a few classes. These include touch (sensitivity to direct contact), vision (sensitivity to light), hearing (sensitivity to vibrations in air, water, or earth), and sensitivities to chemical substances akin to smell and taste. The other possible sense-modalities only occur rarely, for reasons that will be discussed below.

The sensory transduction process is quite similar for all receptors in animals. The stimulus changes the conductance of the receptor membranes, allowing ions (primarily Na^+) to set up an electrical potential. When this reaches a certain level, an action potential is triggered, and a nerve impulse is sent to the central nervous system[11]. The process is thus an electrochemical one. The intensity of the stimulus is registered by the frequency of nerve impulses. Biological sensory information is thus digitized at a very primitive level in the sensory process and works on a principle of frequency modulation. Receptors can also have different rates of adaptation to stimulus durations. A rapidly adapting receptor will only send out impulses at the beginning of a stimulus, and thereafter cease, while a slowly adapting receptor continues to transmit. Once again, the most fundamental aspect of sensation is intensity, and the response of a sensor with respect to stimulus intensity generally follows a power-law (Stevens Law) or logarithmic (Weber-Fechner Law) relationship.

1.5.1 HUMAN SENSES

The human senses and their qualities are summarized in Table 1-4. Human sensation deals with all three classes of physical interactions, although not in the same degree for each. To Class I, force fields, belongs the sense of balance -- we sense the gravitational acceleration acting on us. Further, we have great sensitivity in touch to direct contact pressure with objects. In Class II (propagating objects), smell and taste are both chemical senses, and we perceive fluxes of thermal energy as heat and cold. To Class III (waves) belong both vision and hearing.

Of all human senses, vision is the most sophisticated. It is also the most prominent in terms of information yielded. If we regard each sense as a communications channel, the channel capacity of vision is eight times greater than all the other senses combined[12]. Hence, the phrase "seeing is believing" -- vision is our primary contact with the world, and a common metaphor for knowledge itself.

The eye is a truly amazing sensor. No artificial sensor yet developed can match its combined flexibility, range of sensitivity, and discrimination. The lens focuses a real image of what is observed upon the retina, which is a spatial array of photoreceptors, the sensor elements that convert the impinging light into nerve impulses. These are divided into two kinds: about 6 million cones and about 120 million rods. The cones, concentrated near the center of vision, are sensitive to the three primary colors and are thus responsible for color vision. The more numerous rods are more light-sensitive, but only register intensity, not color. The normal wavelength range of human vision is 390 to 780 nanometers.

Table 1-4. The Human Senses

sense	sense organs	physical interaction sensed	qualities
vision	eyes	light (electromagnetic waves)	colors high spatial resolution
hearing	ears	sound waves	tones high temporal resolution
touch: mechanoreception	receptors in the skin	pressure (direct contact, motion, etc.) active sense	pressure, contact, vibration, tickle varying spatial resolution -- highest in fingertips
touch: thermoreception	receptors in the skin	thermal flux	hot, cold
touch: proprioception	receptors in the muscles and joints	pressure (stretch, etc.)	positions of the limbs
touch: nociception	receptors throughout the body	tissue damage due to mechanical and other causes	pain
smell	receptors in the nasal passages	chemical substances	various scents
taste	taste buds on the tongue	chemical substances	sweet, sour, bitter, salty
balance	semicircular canals of the inner ear	gravity	orientation with regard to forces
general sensations	throughout body	chemical balances	hunger, thirst

The millions of simple receptors that compose the retina give vision great spatial resolution. This makes it pre-eminently our *sense of space*. All our senses have some spatial aspect (even for direct contact), but only in vision and touch are spatial resolution highly developed. It seems certain that the hallmark of reality (for human beings, at any rate) is to simultaneously see something and touch it. To the degree that the world is composed of moving bodies with sizes and shapes, vision is the primary way we apprehend *things*. Further, in mathematics, we always seem to fall back on visual images and models. Vision is also the only human sense that has any useful spatial discrimination at long distances. The binocular feature of human vision gives us depth perception as well, but discrimination in this third dimension is not nearly as good as in the other two.

Hearing is the next most important of the human senses. Sound waves are collected by the outer ear, transferred to the inner ear, and then detected by the tiny hair cells of the cochlea. These hair cells are each tuned to a narrow range of sound frequencies; they function, in other words, as an array of bandpass filters. This gives hearing great resolution among frequencies, and a great sensitivity to the details of waveforms. In vision, there are only three primary colors and we perceive a combination of different wavelengths of light as a single color, blended from the three primaries. In hearing, however, there a whole series of "primary tones" running the gamut from 20 hertz to 20 kilohertz, and a chord always sounds different from a single tone. Moreover, the same musical note played by a piano, a guitar, or a saxophone sounds quite different to us, just as different voices saying the same word can be clearly distinguished.

Compared to vision, the spatial discrimination of human hearing is rather poor. Its immensely better grasp of frequencies, however, makes it the most *temporal* of our senses. We invariably think in auditory symbols: words. And if we grasp things primarily through vision, it is through hearing that we communicate with people. Hearing is the social sense.

Smell and taste are both chemical senses, and are related. Their spatial aspect is very weak. Taste has the four qualities of sweet, sour, bitter, and salty, each with a corresponding region on the tongue. Qualities of human smell are less clear-cut, but seem to be flowery, etheric, musky, camphorous, sweaty, rotten, and pungent[13]. For human beings, smell and taste are essentially auxiliary senses which help further identify things apprehended already by the other senses. Human smell and taste are far less sensitive than their equivalents in many other animals. In dogs, for example, the acuity of smell means that "smelling is believing".

What we normally call the sense of touch actually comprises three distinct modalities. The first is mechanoreception, the sensitivity to contact and pressure of objects with the skin. The mechanoreceptors themselves come in several sorts, detecting pressure, velocity, acceleration, and stretch. These lead to perceived qualities of pressure, contact, vibration, and tickle[14]. Mechanoreception has varying spatial resolution, most acute in the fingertips and at the tip of the tongue, and least acute on the back. Another aspect of touch is sensitivity to thermal conditions. Appropriate thermoreceptors are found in the skin, and these lead to our perceptions of hot and cold. There are specialized nociceptors that detect tissue damage and are responsible for the quality of pain. Finally, there is proprioception, the sensation of the position of the limbs. Proprioceptors are found in the muscles and joints, and provide information about the bend of joints, the stretching of muscles, etc.

The foregoing is basically a summary of the five classical human senses. There are, however, other bodily sensations that do not fall into the classical categories. First of all, there is the sense of balance. The sense of balance is a sense of orientation with regard to the gravitational field, as well as rotational motions. Its sense organs are the semicircular canals of the inner ear. There are also general sensations, pertaining to the entire body. They detect internal chemical balances and are responsible for our sensations of hunger and thirst, for example.

In summary, the human body is equipped with a variety of senses, each fulfilling a different role, none of which alone is entirely adequate to human needs. That is why we regard the loss of a sense modality as a great misfortune. The entire sensor suite, whose information is integrated in the brain, is required for normal human existence.

1.5.2 ANIMAL SENSES

Animal senses, particularly in mammals, are similar to those of human beings. Even when color vision occurs in other vertebrates, it tends to assume a trichromatic form. In some instances, however, animals are equipped with senses quite unlike those we are familiar with. Certain snakes, the rattlesnakes and pythons for example, have the ability to detect infrared radiation. This allows them to find their warm-blooded prey in the dark. The infrared light is not seen with the eyes, but by specialized pit organs in the skin along the snake's jaw. Each pit organ is essentially an infrared pinhole camera. This "infrared vision" is apparently fully integrated with the rest of the snake's visual system in the brain[15].

Many animals have chemical senses far more acute than human smell and taste. We are all familiar with great olfactory sensitivity of dogs, for example. There are animals whose chemical senses go well beyond this. One of the most amazing is the sensitivity of the silkworm moth. It is known, for example, that the antennae of the male silkworm moth can detect a *single* pheromone molecule emitted by the female silkworm moth. (And, perhaps even more astonishing, the female moth cannot smell her own pheromones at all!)[16]

Several species of bats have the ability to navigate and locate prey through ultrasonic sonar. This is a rare instance of a long-range active sense in animals. The bats emit an ultrasonic tone that is reflected by objects, and these echoes are heard by the bats. The bat's hearing is extremely sensitive to the small frequency and amplitude changes that result from the fluttering wings of its insect prey. At the same time, bats are able to correct for the effect of their own flight speed and wing beats[17]. To achieve the sort of spatial accuracy needed to find flying insects, the bat's ultrasonic pulses are in the 80 kilohertz range -- well about the 20 kilohertz maximum for human hearing.

Fishes have an external sense organ known as the lateral line. It was commonly believed that the function of the lateral line was to sense low frequency sound waves, but it now appears clear that its real function is to sense the flow of water around the fish, giving the fish a sense of its motion[18]. There are certain species of fish that have the remarkable ability to communicate and locate objects by electrical pulses. Here is another rare example of a naturally-occurring active sense, although it extends only a meter or two in range. For many species, it appears that the communications function of the electrical pulses is more important than their use for location[19].

Animals may also make use of different aspects of the same sense modality used by humans. Human vision is sensitive to intensity (brightness) and frequency (colors). There are animals -- bees, ants, and certain birds, for example -- that are sensitive to another aspect of light: polarization. Since (due to Rayleigh scattering) the light coming from a clear blue sky is polarized, this allows the animals with such sensitivities to navigate by skylight alone[20].

We note that while the senses of animals and humans cover all three major classes of possible physical interactions, some possible sense modalities to not naturally occur. In particular, long-range active sensation (e.g., sonar, radar) is only weakly developed, although such senses would seem to be of great potential use. There are several possible reasons for the missing sense modes. The first is great difficulty of sensing some physical carriers biologically. Radio waves, for example, require large antennas and complicated circuitry to detect. The second is the lack of biological utility of some possible sense modalities. An organism could easily have a sense for beta or gamma rays, but such a sense would biologically useless. The third, closely related to the second, is the lack of an ambient source of the physical carrier. The eyes of cave fishes have atrophied and vanished, since they live in continual darkness. To see in x-rays might be a great advantage for animals, if only there were a constant source of x-rays around. The range of light wavelengths we see corresponds closely to the optical "window" in the atmosphere -- higher and lower wavelengths are absorbed and do not reach the surface of the earth. Finally, there is the sheer parsimony of Mother Nature -- an organism cannot expend its precious energy on sense-modalities that are useless or redundant. This probably explains the lack of active sensation in most animals. Active senses have typically developed in cases where passive senses, relying on an ambient source of illumination, cannot be used. Echolocation by bats is a clear example of this -- bats use sonar because they hunt at night.

These examples show us that, from a biological point of view, not all sensory modes are created equal. Some stimuli are more easily detected than others: touch is probably the most primordial of the senses for that reason. Stimuli also vary widely in their ability to convey information and in the kinds of information conveyed. Visual light is unique in its ability to provide detailed spatial information about distant objects under ordinary conditions. In contrast, the spatial information conveyed by airborne or waterborne chemical substances is at best vague, but the almost limitless variety of chemical substances means they have great utility as scent markers and in identifying food. It is no accident that higher animals possess the set of senses they do and that other sensory modes have developed only under special conditions.

At the same time, the other possible sensory modes -- although of little use to living beings -- are necessary for a complete scientific view of the world. Not surprisingly, artificial sensors arose largely to remedy the limitations of the human senses.

1.6 ARTIFICIAL SENSORS

Artificial sensors are of relatively recent origin and are not as yet as sophisticated as the sense organs of living beings. It is debatable what the first artificial sensor was, for many measuring instruments could be considered sensors if one stretches the term. If we return, however, to the sensor's fundamental purpose of transforming physical energy into information, the first artificial sensor was probably the magnetic compass. A few centuries later, it was followed by the thermometer. From that point onward, the development of sensors and the instruments of experimental science was closely related. Artificial sensors remained rudimentary until the development of electronic equipment in the 20th Century. Electronics, followed by digital computer technology, provided a means of reading out and representing complex artificial sensory information. Modern artificial sensors, with few exceptions, produce an electrical representation of information, whether it be in digital or analog form.

Sensor technology in the 20th Century has advanced dramatically. The numbers and types of artificial sensors is so vast that it can only be summarized here. For more details, I refer the reader to some standard references[21]. Tables 1-5 through 1-7 summarize artificial sensors organized by modality.

Artificial sensors tend to be most used to provide exactly that information which human sensation cannot. Artificial sensors allow us to detect objects in the infrared or by electric fields, for example. They also provide active modes of sensing, such as radar, ultrasound, and x-rays. Artificial sensors can go into environments where human beings cannot, and they provide continuous monitoring of locations (remote sensing) where it would be prohibitively expensive to maintain human surveillance. Perhaps the best example are unmanned space probes. The Viking landers on Mars, the Mariner and Venera probes to Venus, and, most recently, the Galileo vehicle that plunged into the atmosphere of Jupiter, were all equipped with arrays of sensors to relay scientific information about the composition of atmospheres and soils. Most probes also carry cameras to provide a close-up picture.

The most important question raised by sensor theory in this area is: do artificial sensors represent an extension of human knowledge? The answer to this question will be in the affirmative -- an artificial sensor, according to sensor theory, is no different from a biological sense organ. Human beings have thus devised ways of extending their sensation to modes previously inaccessible, detecting all physical interactions directly, except the microscopic primary forces and waves. This is a point of some epistemological importance, although it has been almost totally ignored by philosophers. Sensation is not simply a problem of physiology or psychology or cognitive theory. It is one of the relation of the knower to the object known, and that relation can be through an artificial sensor no less than the five human external senses. Artificial sensors have brought our minds into more intimate contact with the world than was previously possible.

Table 1-5. Artificial Sensors of Force Fields.

physical interaction	artificial sensors
electric fields	capacitors
magnetic fields	Hall effect sensors, magnetoresistive sensors
gravity	pendulums
microscopic primary and secondary forces	(inferred, not detected directly)
fluid pressure	mercury pressure sensors, membranes, bellows, piezoresistance sensors
solid pressure and strain	strain gauges, tactile sensors, piezoelectrics

Table 1-6. Artificial Sensors for Propagating Objects.

physical interaction	artificial sensors
electric current	ammeters, voltmeters, etc.
radiation: elementary particles and nuclei	geiger counters, scintillation counters, cloud chamber, spark chamber, cerenkov detector, etc.
atoms, molecules, and chemical substances	mass spectrometers, electrochemical sensors, enzyme and catalytic sensors, humidity sensors.
fluid flow	pressure gradient sensors, thermal transport sensors, ultrasound.
motion of macroscopic objects	in an object: accelerometers, electromagnetic velocity sensors. outside an object: ultrasound, radar, infrared cameras.

Table 1-7. Artificial Sensors for Propagating Waves.

physical interaction	artificial sensors
radiofrequency (RF) and microwaves	radio receivers, radiotelescopes and imaging radar.
infrared light	specialized phototransistors, photodiodes (e.g., InSb, HgCdTe, PtSi, etc.) cooled detectors.
visible and ultraviolet light	television cameras, photodiodes, phototransistors, photoresistors, photomultiplier tubes.
x-rays	semiconductors.
gamma rays	semiconductors, other elementary particle detectors.
other primary waves (weak, strong)	(inferred, not detected directly)
sound waves	resistive, condenser, fiber-optic, piezoelectric, and electret microphones.

CHAPTER TWO

PROTOSENSORS AND SIMPLE SENSORS

2.1 THE MEANING OF SENSORY TRANSDUCTION

Let us examine in more detail exactly what occurs when a sensor responds to a stimulus. Every sensory stimulus, to repeat, has some energy intensity associated with it. Without energy, the sensor cannot be affected and sensation cannot take place. Thus, the primary thing that happens in transduction is the capture of the stimulus energy by the sensor, causing it to produce an output. Since the sensor is a physical object, this output is also energy. In both sense organs and artificial sensors, the sensor output is usually an electrical signal, regardless of what kind of stimulus the sensor is designed for. But it need not be energy of the same type as the stimulus; the new carrier of information in theory could be anything. Transduction thus completely converts the kind of energy involved. There is no necessary connection between the energy types of the stimulus and the output. That is, one could not infer the mode of stimulus energy *merely* from the output energy. (Energy is thus the "matter of sensation" -- it is the carrier of patterns and information, but impenetrable in itself.) We do know, however, that energy is present in both the input and output of the sensor.

The energy intensity is the *quantity* or magnitude of the sensation. In transduction, this magnitude is converted to a symbolic representation. This can be a direct or proportional conversion, as when, for example, an intensity of light is transduced into an electrical current. It can also be indirect: an intensity can be given a digital representation, such as a binary number or a frequency of discrete nerve impulses. It depends on the internal representation of the information system. The important point is that the level of intensity is given a symbolic form on the new (internal) energy carrier. The stimulus intensity becomes another kind of intensity or magnitude.

Stimulus intensities can have patterns in space and/or time. These patterns are often what sensors seek most of all to ascertain. They may be symbolized directly (an array of numbers corresponding to a spatial field of intensities) or indirectly (a given pattern is represented by a single symbol, like "triangle" or "circle"). The spatio-temporal pattern is not transduced. It is either transposed from the old energy carrier to the new one, or else it is recognized and denoted by a symbol. The information has a new carrier. In either case, the spatio-temporal pattern -- in contrast to the energy of the sensation -- has actually been conveyed into the information system. There is a relation of the two. This also occurs in the transduction of the state parameters of stimulus. These are symbolized as intensities are: light frequency is translated into color, for example. Their order also conforms to the order of the stimulus, as will be shown. The important point is that these patterns are the *qualities* of sensation. They are also the "form of sensation", in that they are the ordering principles of the sensation. In transduction, the ordering principle is made to act on new elements: the form is imposed on a new matter, so to speak. Consider a visual image projected on the retina. The image's spatial pattern becomes an analogous spatial pattern of impulses in the optic nerve. The image's colors at any point are symbolized by excitation of certain nerve fibers and not others.

In transduction, then, the energy associated with the sensation is completely converted. The patterns of the energy in intensity level, space and time, and state parameters are not converted, but translated. This raises a couple of questions for sensor theory. What is the simplest imaginable sensor that can carry out transduction? And, of even more importance, can other sensors be represented as a combination of these simplest sensors, as molecules are built from atoms?

2.2 THE PROTOSENSOR AND ITS CHARACTERISTICS

Let us call the simplest imaginable sensor a *protosensor*. A protosensor is the most elementary device that can perform the essential function of a sensor: the conversion of energy into information. To understand its characteristics, we must consider the protosensor from both the physical and the informational sides of the interface.

On the informational side, the protosensor has one characteristic. Its output is a single binary bit. Whenever the protosensor's stimulus is present, it outputs "1" ("on" or "true"); when the stimulus is absent, it outputs "0" ("off" or "false"). The protosensor is thus like a switch that can be either on or off. Its output tells us simply if its specific stimulus is present or absent.

On the physical side, the characteristics of a protosensor are necessarily more numerous. A "bare", "unconditioned" protosensor would respond to any stimulus, energy of any kind. However, it must be possible for a sensor -- including a protosensor -- to operate in our physical world, where energy and possible physical interactions are not generic, but occur in well-defined types. Further, to be useful in the description of sensors, the protosensor must be an infinitesimal sensor, an idealized point. This precludes a general sensitivity to all stimuli. It requires the opposite: sensitivity to a *single* mode of stimulus. This means that protosensors, no less than real sensors, have certain conditions imposed on them by physical laws.

The protosensor has five conditions or limitations imposed on it by physical reality: mode, method, band, field, and range. For the protosensor, these are infinitesimal, but for simple sensors (to be discussed below), they become finite. These properties are also constant in time.

The *mode* of a protosensor is defined by the kind of physical energy it detects. The different kinds of physical interactions were enumerated and discussed in the previous chapter. We saw that the modes of physical interactions correspond to the possible modes of sensing. A protosensor has a single mode, which is to say it can sense a single kind of physical stimulus.

For any sensory mode, there may be several possible *methods* of detecting the physical carrier of the interaction. For example, the sensor can measure an instantaneous flux or a time-integrated fluence; it can count the number of impinging objects or measure their energy, and so on. The method of sensing is what provides the *units* of the sensor output.

The first chapter also emphasized, that for each possible mode of sensation, there were several parameters that defined the state of the physical carrier. Light, to recall, has a frequency and polarization. These parameters are sub-modalities or qualities. A protosensor is limited to a *single* value for each parameter: a single wavelength of light or a single type of molecule. This defines the *band* of the sensor.

Next, we must consider the fact that the sensor is viewing objects that exist in space and time. Moreover, the protosensor must itself be considered a physical object. A protosensor is limited to an infinitesimal point in space; it has a spatial extent of *dx dy dz*. This extent defines the *field* of view of the sensor. The sensor collects energy -- responds to its stimulus -- only within its spatial field of view. Stimuli outside that field are either ignored or not received.

Finally, there is the *range* of intensities a sensor can respond to. Occasionally, this is called the *span*, input full scale, or dynamic range of the sensor[1]. For a protosensor, the range is a single intensity level, defined in terms of units of energy or number, depending on the method of detection. It can be that the intensity is directly coupled with the band, as in the measurement of temperature.

A simple example of a protosensor would be a sensor that registers "1" whenever it receives light (mode), detected as an energy flux (method), with a wavelength of 550 nanometers and right-handed circular polarization (band), at a single point on the earth's surface (field), and with an intensity of 10^{-6} watts/cm^2 (range). The output of any protosensor means that a stimulus matching its characteristics is present -- a binary quantity attached to a quality, in other words. The meaning of a protosensor's output is thus derived from the five characteristics or physical conditions.

Very few actual sensors can be approximated by a single protosensor. The human sense that comes closest in this regard is touch, in its aspect of sensing direct contact with objects. We sense such contact in a binary, yes or no, fashion. Some of the very fast mechanoreceptors indeed will send out only a single pulse to indicate contact has been made. Although the photoreceptors of the eye are not close to being protosensors, it is notable that light and dark are encoded in a binary way, along either on-center neurons (registering the presence of light) and off-center neurons (registering the absence of light)[2]. We could imagine an artificial retina that consisted of a vast spatial array of protosensors, reporting information in precisely the way the human optical system does.

The importance of the protosensor does not rest in a similarity to any actual sensor. Its importance derives from the opposite fact that it is an idealization. The protosensor is to sensors what the infinitesimal lines and points of geometry are to physics. Mathematical laws and mathematical idealizations allow the physicist to decompose the motion of an object into velocity and acceleration vectors. The whole success of mathematical physics is based on this fact. The protosensor offers the same possibilities for sensor theory. It is an idealization, but any sensor may be decomposed into protosensors as motions are decomposed into vectors. The protosensor is the elementary theoretical building block of all sensors. Since any sensor can be represented as a combination of protosensors, the information output of any sensor can likewise be represented as a binary number corresponding to the outputs of its component protosensors.

The details of such combinations will be considered at length in the next chapter. Here it is enough to say that every sensor, just like a protosensor, can be characterized by mode, method, band, field, and range. The decomposition of a sensor into protosensors expresses the fact that actual sensors are not limited to an infinitesimal value in each characteristic. They may cover a finite (even infinite) extent in each, with a certain number of increments or levels. If there is only one increment for each characteristic, then we have a *simple sensor* or sensor element, to be discussed in section 2.4 below. But now we must consider the important problem of the reliability of the information output of a protosensor.

2.3 PROTOSENSORS AND KNOWLEDGE

The question of the truth of protosensors is the central epistemological point of sensor theory. The reason for this is quite simple. All sensors can be represented as a combination of protosensors. If protosensors are not reliable, then no actual sensor can be, because a sensor can be only be as true as the protosensors that virtually or actually compose it. Therefore, the possibility of a protosensor registering truthfully is the same thing as the possibility of sensation as a source of knowledge. If protosensors are reliable, then knowledge of the sensible world is not only possible, but necessary, on the theoretical level, and is practically possible to some limited degree. If not, then such knowledge is impossible, and sensor theory (as well as sensation) is of no philosophical interest -- it is little more than an engineering problem.

Even at the purely practical level, the question of the reliability of the protosensors is important. If protosensors -- an idealization -- cannot register truthfully, then this makes understanding the reliability and errors of actual sensors very problematic. Ultimately, the distinction between signal and noise must rest on this.

As a first step, let us review the meaning of the information yielded by a protosensor. Whenever the output state of a protosensor is "1" ("yes", "on", "true"), it means that there is something present with characteristics typical of the protosensor. We know that:

1) a stimulus is present and acting on the protosensor.
2) the type of stimulus (physical interaction) corresponds to the sensor mode.
3) the stimulus has state parameters (e.g., frequency and polarization) corresponding to the sensor band.
4) it is located at a point in space corresponding to the sensor field of view.
5) it has a certain energy intensity, corresponding to the sensor intensity range, with units defined by the method of sensing.
6) the stimulus originated from an object, either by emission or reflection. An actual relation or connection thus exists between the sensor and an object.
7) the object has specific properties that allowed it to emit or reflect that stimulus (physical interaction).

Conversely, when the protosensor reads "0" ("no", "off", "false"), that means that nothing is present with the above characteristics. Either no stimulus is acting, or it is there, but lacks the sort of properties -- either of mode, band, or location -- the protosensor can detect. There is no vagueness here: something is either present or absent corresponding to the protosensor's characteristics. That is because, once more, the protosensor is an ideal and perfect sensor.

The output of a protosensor defines, if you like, a virtual proposition that: something in the world = the protosensor's conditions. This is also the measure of the validity of the protosensor: if the protosensor is "on" only when its conditions are met, then it is reliable and valid. This may tell us very little about the object, but it is key to the whole problem of sensory knowledge. Since all sensory information can be decomposed into protosensory information, it is only as reliable as the protosensors themselves.

This brings us to the main point regarding the validity of protosensors. Not only is a protosensor exact, *it cannot be in error.* It always reports truthfully and cannot be tricked. This follows from the definition of the protosensor. Whenever a protosensor registers "1", its conditions are met, and something in the world with those characteristics is present. It is *necessarily* true. The absolute validity of protosensors has far-reaching implications which must be both thoroughly justified and explored.

It must be emphasized once more that the protosensor is an idealization -- it cannot malfunction in a *practical* sense. We are interested here in investigating if a sensor can be valid even in theory. If sensation cannot attain complete validity in an idealized case, it certainly cannot do so for real sensors. For sensor theory, errors in *actual* sensors arise because they are not idealized protosensors or a perfect combination of protosensors.

Every sensation -- every sensory data representation -- can be built-up from protosensor outputs. It can now be shown that the validity of protosensors is a condition of the possibility of any sensation. That is, protosensors must be absolutely reliable and true, or sensation itself is impossible. And the self-evident fact that we do sense thus implies both the absolute validity of the protosensor *and* the potential validity of actual sensors.

There are a couple of approaches to show the connection between protosensor validity and operability. Let us assume for a moment that protosensors were essentially unreliable and false. (Not a malfunction: protosensors are idealizations.) There are three ways in which such unreliability could occur:

1) The first way is if the protosensor failed to respond to its specific stimulus; i.e., that it registered "0" when it should have registered "1". But if this were the case, then the protosensor could respond to no stimuli at all; it would be "dead" and no sensation would be possible. A protosensor is like a tuning fork: it tuned to one "note" only. If it does not function for its specific stimulus, it cannot function at all.

2) The second way is if the protosensor responded to stimuli that were outside its specified conditions: that is, responded when its specific stimulus was absent. If this were the case, however, the protosensor would always register "1", would always be "on". No distinction is made between the presence or absence of the stimulus. Thus, the protosensor's output would be completely meaningless, and -- as in the case of the "dead" protosensor -- sensation would be impossible. (Note that if the protosensor conditions were different from what we thought they were -- e.g., tuned to Ab rather than A -- this is an error in our knowledge of the conditions, not of the protosensor itself. The protosensor will still respond to a stimulus corresponding exactly to its conditions.)

3) The third way would be if the protosensor responded truly most of the time, but failed at others -- that it were 98% reliable, for instance. This argument springs readily to the practical mind. However, for just that reason, it is an attempt to impose practical conditions on an idealized situation. The fact that the circles we attempt to draw on a chalkboard are not perfectly round does not cancel out the fact that the circumference of an ideal circle equals 2B times the radius. Likewise, a protosensor is perfect and has constant properties. If it registers untruly, it must be *essentially* unreliable, not just an occasional malfunction. A dynamically unreliable protosensor could only be one that randomly and unpredictably registered true and false. But in this case the protosensor output could only be a random noise, with no distinction drawn between stimulus and non-stimulus. The data representations produced would thus be entirely disconnected, objectless, and meaningless. Sensation would once more be impossible.

In each case, we see that if protosensors are unreliable, sensation is impossible. The absolute validity of protosensors is thus a condition for the possibility of sensing. If sensation is possible, then protosensors are always valid. This can be shown in another way. Assume that sensation is objectively possible. What, from the point of view of sensor theory, makes it possible? First, if sensation occurs, an object is causing data representations within an information system. This requires the intermediary of a sensor. We have already seen that any sensor can be decomposed into protosensors and any sensor output into protosensor outputs. Second, for a protosensor output (a "protosensation") to occur, there must be a stimulus present that meets the protosensor's conditions, because a protosensor will only respond to such a stimulus. Since the protosensor only responds to stimuli that meet its conditions, sensation will become possible only if the protosensor is valid.

One possible objection is: how do we know the stimulus originated from an object? Why must the stimulus put the sensor in relation to an object; why can't there just be a free-floating cloud of objectless stimuli? First of all, the sensor stimulus must be physical. A sensor, by definition, receives and responds to only physical stimuli. Second, since the stimulus is a physical interaction, it must come from something with a physical manifestation. If there were a free-floating cloud of stimuli, it must have originated from some physical object or event, even if that object is the universe as a whole[3]. The physical interactions that make sensation possible are just that: interactions, involving two or more physical things. Thus, the detection of a stimulus must, of its nature, means that a relation or connection exists between an object and the sensor, however tenuous or remote it might be. Sensation requires relation to an object, and it also requires that a distinction be drawn between the presence and absence of the stimulus. (We will consider the inference of the *properties* of the object in later chapters; here, only the existence of the relation to the object is established.)

Finally, let us take our critical exploration to the extreme limit. Could an object or even something like Descartes' "evil genius" cause false sensations or a false reporting of information? A false reporting of information would be possible in two ways: (1) malfunction of the protosensor or (2) bypassing the sensors and acting directly on the information system itself. As we have seen, a protosensor cannot malfunction; it is necessarily true. Thus, false reporting of information is not possible in that way. Even if physical stimuli could be streamed at the sensor *ex nihilo*, they would still be coming from a physically manifested event. Likewise, there is no possibility of a protosensor manufacturing its own object. The false objects sometimes found in actual sensors, like afterimages in the eye, are due to the limitations of sensor materials and construction. In the second case, the bypassing of the sensor, a false report is indeed possible. But in that case, what we have is no longer sensation. The whole purpose of a sensor is to act as an interface between the physical world and an information system. If we plug ourselves directly into the information system, we may introduce whatever falsehoods we like, but this in no way impugns the reliability of the protosensor. Moreover, a simple intervention into the information system can be defeated by active sensing, a fact discussed at length in Chapter 9.

In summary, we see that the validity of protosensors is directly connected to the possibility of sensation itself: the protosensor will respond only when there is a correspondence to the characteristics of the object. Sensation requires a match between the sensor and the object. It is a mutual conformation, like a key and a lock. When the conditions of an object and a sensor match, the object can be sensed, and hence *become an object for that sensor*. The object thus conforms to the sensor, because if it does not match the sensor's characteristics, it will never be detected by that sensor. It thus will never become an object of experience for the information system employing the sensor. And we will only "see" those aspects of the object that the sensor characteristics allow us to see. At the same time, however, the sensor is telling us something real about the object. The sensor conforms to the object for this reason. Sensation (as considered from the point of view of a perfect protosensor) requires this complementarity. When the object and the sensor match -- can interact in the same way with the same physical carrier -- then a relation of the two, and hence sensation, becomes possible. If not, then no relation is possible, and thus no sensation or experience of the object. We will return to this analogy of key and lock for sensation many times in the course of this work. A protosensor, of its very nature, conveys a tiny piece of information about a physical object. Later chapters will explore what this information is, and how we can cross over from the sensory appearance of an object to its nature.

2.4 THE SIMPLE SENSOR

All protosensors are infinitesimal *simple sensors*. Conversely, simple sensors are exactly like protosensors, except that they have a wider extent in band, field, and range. Like protosensors, all simple sensors are limited to a single mode of sensing and a single method. Further, they still produce a binary output.

Simple sensors are often called *sensor elements*, *detectors*, or *receptors*. These terms express the fact that there are actual sensors that come close to being simple sensors (as will be discussed below), even though the simple sensor itself remains an idealization.

The expansion of the band, field, and range to finite (or even infinite) values means that simple sensors have several new characteristics beyond those of the protosensor. For example, for the protosensor, the band consisted of a single point -- a single wavelength and polarization of light, for example. A simple sensor must define limits for each of these parameters of the physical carrier -- the simple photodetector, for example, has low and high wavelengths of light to which it is sensitive. The band (or bandwidth) includes those wavelengths between the limits.

The sensor field of view is likewise expanded in the three spatial dimensions. The sensor is no longer limited to a single point, but can cover a one, two, or three dimensional region. It will also have an orientation with respect to other objects. Since a sensor measuring a flux will usually be two-dimensional, the orientation will define a line-of-sight viewed by the sensor. The sensor field of view can also be an entire physical object to which the sensor is directly connected. A good example of this is a thermometer measuring the temperature of an engine or of a person.

Perhaps the most important expansion is that of the range of intensities. Even the simplest actual sensors have a finite range of intensities. The minimum value of intensity is the sensitivity *threshold*. Below threshold intensity, the sensor does not respond to the stimulus. At the other limit, sensors often seem to have no upper limit to the intensities to which they will respond. (Complex sensors, as will be discussed, can have a saturation point.) However, above a certain energy intensity, any sensor will be physically damaged or destroyed. Thus, there will always be a practical upper limit for intensities, though it may be quite large.

The information produced by a simple sensor is just like that of the protosensor: it is a binary number. If the sensor output state is "1", then the simple sensor's conditions are met and a stimulus is present with its characteristics; if not, then the output state is "0". The meaning of the output is more vague than for the protosensor, because the simple sensor collects any energy that falls into its finite "bin". The information yielded is just as reliable as that of a protosensor, however. Consider an optical sensor that detects all visible wavelengths of light. Whenever it registers "1", that means that something visible is in its field of view.

There are many kinds of actual sensors whose characteristics approximate a simple sensor. The photodetectors that switch streetlights on at dusk and off at dawn are close to being simple sensors. They are sensitive to a wide band of wavelengths of visible light and cover a hemisphere in view. They have a defined threshold above which the lights are switched off -- the output of the photodetector is binary, in other words. Another example is a thermostat. Above or below its threshold (the desired temperature), it signals the furnace or the air conditioner to start or stop. Indeed, most sensors that act as binary switches of this sort can be approximated as simple sensors. Regardless of the details of the sensor, the information output has been limited to a single bit which is toggled when a threshold condition is met.

Human sense receptors also function in a way close to simple sensors, if we consider them instantaneously or over a very short time interval. This is due to the fact that sensory information is reported as a series of binary nerve impulses, with the frequency of pulses indicating the intensity of the stimulus. The receptor is like a bucket that when filled is suddenly dumped and sends out a binary signal to indicate that the dump has occurred. Each receptor thus has both a threshold intensity and a maximum intensity (the "filled bucket").

If the simple sensor's extent -- in band, spatial field of view, or intensity range -- is wide, then the information it yields is vague and general. On the other hand, if the extent is narrow, the information is specific, but likewise limited. To obtain information which is both wide, yet detailed, requires an effective combination of many sensor elements. The use of the term "sensor element" for simple sensors indicates they are the real building blocks of complex sensors. The ways of combining simple sensors to form complex ones will be treated in the next chapter. As a preliminary to that discussion, we must note there are two possible ways of limiting the band, field, and range of a simple sensor. The first is the ordinary way of bounding, a sharp cut-off. For example, consider the band for a photodetector: the wavelengths λ of light which the sensor can detect fall within the band limits, $\lambda_1 \leq \lambda \leq \lambda_2$. Wavelengths outside of this range are ignored. With such ordinarily bounded simple sensors, we will build up complex sensors like tiles on a floor, or segments in a line.

There is, however, a second way of limiting simple sensors in band, field, and range. And that is, instead of having a sharp cut-off or boundary, to suppose they have a Gaussian form. To reconsider the photodetector in the previous paragraph with Gaussian limits, the response of the detector would fall off gradually from a central wavelength λ_c proportional to $e^{-(\lambda - \lambda^c)^2}$. This leads to a "smooth" limit for the sensor element, rather than a sharp cut-off.

The discrete, sharply bounded, way of viewing sensor elements is the normal and most natural approach. Indeed, complex sensors are often constructed in their spatial layout from simple sensors like tiles on a floor. The Gaussian element approach has several advantages, however. First, the Fourier transform of a Gaussian is another Gaussian, while the Fourier transform of a bounded element is the sine integral function (sinc). Second, the use of a Gaussian element allows one to easily decompose a continuity into virtual sensor elements, because the errors of any actual sensor will assume a normal (Gaussian) distribution. Taking the standard deviation of the error distribution, we obtain an effective element size.

CHAPTER THREE

COMPLEX SENSORS

3.1 KINDS OF MULTIELEMENT SENSOR

Few actual sensors, as we have seen, can be represented by a single protosensor, which is essentially an ideal limiting case. Nor do most sensors, sense organs, and sensory systems consist of a one simple sensor. They integrate many simple sensors, often in vast number, to form a *multielement sensor*. This is certainly true of sense organs: the eye has millions of rods and cones, the hands have many thousands of mechanoreceptors, and so on. Most artificial sensors are also multielement. An infrared optical sensor will often have hundreds of sensor elements on its focal plane. A radar will have a great sensitivity to subtle changes in frequency: in effect a large number of band elements.

The aim of any multielement sensor is to cover a larger band, field of view, or intensity range, or to obtain a wider view of the world through different modes or methods of sensing. While the multielement sensor covers a wider extent, it also maintains the finer resolution of the simple sensors that compose it. A single sensor element or protosensor yields information that is exact, but narrow. A simple sensor covering a wide range is necessarily vague. A multielement sensor tries to combine both breadth and accuracy of information about the world. Not surprisingly, the output of a multielement sensor is not a single binary bit, but an binary word uniting the information from each sensor element.

Even something as basic as several sensor intensity levels requires a corresponding number of simple sensors -- one for each level. A continuum, an analog output, can be represented as a combination of digital simple sensor data. That does not mean that all sensors are actually decomposed into simple sensors, just that they can be represented in that way. The decomposition can be virtual, in other words, just as a continuous line can be broken into virtual parts.

The next step in sensor theory is thus to find the rules for representing multielement sensors in terms of simple sensors. This is the problem of *sensor fusion*. The goal of sensor fusion is to take the data outputs from two or more sensors and synthesize them into a single representation. The multielement sensor output is formed from a combination of binary simple sensor information. Sensor fusion requires a commonality or likeness between the sensors as an underlying condition of the data synthesis. To find the possible commonalities and rules for data synthesis is the whole sensor fusion problem.

The most important division between sensors is the mode of sensing, as was described in Chapter One. Different modes may require dramatically different kinds of sensing methods. Consider how different the eye is from the ear, for example. There is a corresponding division between kinds of sensor fusion -- synthesis of sensor elements belonging to the same mode or synthesis of sensors belonging to different modes.

The combination of sensor elements of the *same* mode leads to a *complex sensor*. Complex sensors contain two or more sensor elements specialized as to band, field, or range. The unification of the elements will be according to the shared commonality of mode, and similarity in the band, field, or range. The human external sense organs are all complex sensors.

The combination of sensors -- whether simple or complex -- of different modes or sensing methods leads to a *multimode sensor* or *sensor suite*. Multimode sensor fusion is treated in the next chapter. The unification of different sensor modes must be based on something belonging to the object or shared by all the sensors, usually of a numerical or geometric nature. Since the problem of multimode sensor fusion is more difficult and usually includes complex sensors, the characteristics of complex sensors must be examined first.

3.2 COMPLEX SENSOR SYNTHESIS

The elements of a complex sensor, to repeat, must belong to the same mode. The shared mode is the basic commonality that enables data synthesis and gives the complex sensor output meaning. Consider two simple optical sensors, for example, with different bands, fields of view, or intensity ranges. The fact that they both detect light permits the fusion of their outputs into a single image.

The protosensor conditions provide the measure of the similarity or dissimilarity of the sensor elements being integrated. These are:

1) the *band* of different states within the same mode.
2) the spatial *field* of view of each sensor element
3) the *range* of intensities which each sensor element can detect.

Band, field, and range define a parameter space in which sensor elements are placed. The number of dimensions of the parameter space is mode-dependent. In an optical sensor, for example, there are the three spatial dimensions, plus the frequency and polarization of light -- four linear dimensions and one circular one. The parameter space thus corresponds to the possible state vectors of the physical interaction.

The arrangement of the sensor elements in the parameter space is the *order* or *form* of the complex sensor. The sensor form is a rule of synthesis provided by the sensor designer (whether a human being or nature), and governs the placement of sensor elements in relation to each other. It must thus also govern the synthesis of the sensor outputs into a single data representation. It is an *a priori* form, intrinsic to the sensor itself, preceding any actual sensation. The "biggest" possible sensor form would be the parameter space itself.

This is much less abstract or mysterious than it sounds. The form or relational rule is generally quasi-spatial in character. Sensor elements can be arranged in a straight line, along a curve, in a rectangular grid, radially, or in any other geometry. Arrangement here is not only in space, but also in the band.

 Rectangular tiling of sensor elements

   Separated sensor elements

 Radial arrangement

 Concentric arrangement

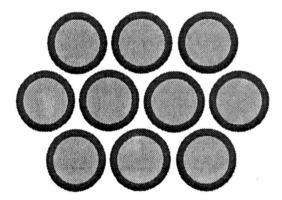 Hexagonal arrangement of circular elements

Figure 3-1. Spatial Arrangements of Sensor Elements

Different sensor elements are united to cover a wavelength band, like the three kinds of cones in the human eye. The sensor elements are most commonly adjacent (like tiles), but can also overlap or be widely separated (Figure 3-1). To return to the example of color vision, the wavelength response curves of the three kinds of cones overlap to form a continuum.

3.2.1 INFORMATION SYNTHESIS OF COMPLEX SENSOR DATA

Closely related to the ordering of elements in a complex sensor is the problem of how to combine the information outputs into a single data representation. The maximum amount of information results from a direct combination of the binary outputs of n simple sensors into a n-bit binary number. This is a *multivalent* synthesis, which permits detection of every possible bit pattern yielded by the sensor elements (2^n in all).

Multivalent synthesis is not the only way of forming a data representation, however. One of the most important kinds of synthesis is the *cumulative*. Here we are interested only in the highest or lowest sensor element -- a magnitude. Intensities are registered in this way. At a given time, there is only one intensity at a point, not a spectrum or pattern of intensities. What is of interest is a vertical series of levels. Sensor elements in a cumulative synthesis are like the level markings on a measuring cup. Measurement of the distance between two objects or detection of the edges of an object are other examples. The information yield of a cumulative synthesis of n elements is $\log_2(n)$ bits. ($\log_2$ is the base-2 logarithm and is equivalent to $3.32 \log_{10}$.) If one is only interested in a magnitude, the savings in bits between a cumulative and multivalent data synthesis are considerable and rise quickly with the number of sensor elements.

An *additive* synthesis sums the outputs from two or more sensor elements to obtain a single intensity or binary output. An additive synthesis of n simple sensors will yield $\log_2(n)$ bits, just as for the cumulative. Additive synthesis occurs in dark-adapted vision[1]. The outputs from several cones are summed, and if the sum exceeds a threshold, a binary impulse is sent to the central nervous system. In many areas of human sensation, this kind of summing and comparing of sense data occurs, often very early in the sensory process.

The simplest possible synthesis of elements is the *monovalent* or *binary*. Its information yield, like that of a simple sensor, is one bit. In a binary synthesis, the simple sensor binary outputs are combined according to either a logical OR or a logical AND. The use of OR effectively leads to the combination of the several simple sensors into a new larger simple sensor. The use of AND registers if all the simple sensors are simultaneously "on"; this results in a simple sensor with a higher threshold.

3.2.2 ANALOG VERSUS DIGITAL

In the "real world", intensities and the carrier parameters (such as wavelength) assume continuous values. Moreover, sensor elements will respond to a continuity of values, especially for intensities. It is analog, in other words, with infinitesimal resolution. To represent such a continuity in the ideal case would require an infinite number of gradations, and the sensor output would consist of an infinite-length binary word -- an infinite amount of information. In all actual sensors, however, there will always be limitations that allow us to represent an analog signal effectively by a finite digital number. The imperfections of materials will always introduce an error (if only through finite time response), and the error defines an effective discrete element size: number of virtual elements = total extent / error. A 3% error in intensity, for example, means we can represent a full range of intensities by 33 levels. In human vision, the limitations of the retina and the nervous system explain why, under ordinary lighting conditions, human beings can perceive only 30 to 40 different shades of gray along a scale from deep black to bright white[2]. Since no actual sensor is perfect, its analog output can be digitized with no effective loss of information.

The fact that an information system can accept only a certain amount of data also limits the number of sensor gradations. Every sensory representation must pass through a communications channel with finite capacity. And the information can be useful without an infinite number of gradations. To return to the example of vision, although the human eye can distinguish millions of different colors in theory, perceptually, the average person can distinguish perhaps 180. A rainbow contains an infinite number of different wavelengths, but we do not see this: the eye and mind break it up into wide bands of distinct colors[3].

Even with the possibility of energy loss, there are some strong advantages to digitizing analog data. There is less chance of data corruption in a set of finite symbols than in an analog continuity. This is probably one of the reasons why sensory information in animals is digitized in the form of nerve impulses.

3.2.3 SYNTHESIS OF INTENSITY LEVELS

The most fundamental kind of multielement sensor fusion is that of intensity levels, and it invariably occurs within each actual sensor element. All sensation has some intensity associated with it. For simple sensors, we defined a minimum intensity to which the sensor responds (the threshold) and a maximum level. A multielement fusion for intensity results in several intensity levels within a wider range. Since most actual sensors respond in an analog fashion to intensities, the levels will be virtual, and the resolution will be defined by the sensor errors, digitization process, and so on.

The synthesis of intensity levels is cumulative, in that different intensity levels are "stacked" on top of each other in an ascending series. Intensity is a magnitude, and only the highest level attained is of interest. The information output of a single sensor element with n intensity levels thus can be represented by a binary number of $\log_2(n)$ bits.

The relation between stimulus intensity and sensor response need not be linear, however -- which is to say that the simple sensors for each intensity level need not cover equal increments. To the contrary, it is very common for sensors to respond in a logarithmic or power-law fashion to the stimulus. This permits the sensor to cover a wide range of intensities, but to maintain a fine resolution for small intensities.

3.2.4 SYNTHESIS OF SPATIAL FIELDS

The next most important synthesis of sensor elements is over space, the combining of two or more sensors at different locations to cover a larger field of view. This is usually achieved through an array of contiguous or nearly contiguous discrete sensor elements. This is true both of the human retina and human touch. Generally, for any sensor where the main interest is the spatial properties of the object, there will be a spatial array of sensor elements, and the most common form this takes is a rectangular grid or tiling. The array of elements need not be regular, but may be more concentrated in one area than in another. In the human eye, photoreceptors have the greatest density in the visual center of the retina (*fovea centralis*), giving this region of vision the most spatial acuity. Mechanoreceptors are most concentrated in the fingertips, giving them the ability to detect fine patterns and small objects.

Even where the sensor elements do not form a regular grid, the information obtained from the sensor is commonly used to create one. One example is scanning. Most television cameras do not have a spatial grid of sensor elements; rather, they have a beam that scans the image in time. The same is true of most radars, whether a phased-array or a mechanical dish -- the radar scans in time to create the spatial image.

Non-contiguous elements may provide information not available to a regular spatial grid of elements. Widely spaced sensor elements will provide one with better discrimination of small distances, at the sacrifice of lower spatial frequencies. Such techniques have been used in radio astronomy (e.g., the Very Large Array) to view distant galaxies and in visual astronomy to view the surfaces of stars.

Separated sensors or sensor elements in two dimensions can also provide depth perception, by having different perspectives. The most obvious example is that of human binocular vision, even though the perception of the dimension of depth is far less acute than for the other two dimensions. A good artificial example is a Computer Aided Tomography (CAT) scan. A medical patient is scanned with a series of one-dimensional x-ray beams in horizontal and vertical directions. The resulting information is stored on a computer and used to reconstruct a full three-dimensional image.

The details of providing spatial images to sensors, and the trade-offs involved in the quality of these images, are the province of optics and need not detain us further. The main point is that the spatiality of a given sensor will depend, not surprisingly, on the number of sensor elements it has spatially arrayed. There must be a good proportion between the spatial concentration of sensor elements and the spatial detail of the objects which the sensor views. The spatial acuity of human vision and touch are great as they are our main contact with the locations, shapes, and motion of things. The spatial acuity of hearing is much less, and that of smell and taste is vague indeed.

Spatial data synthesis is most typically multivalent. The spatial information from each sensor element is preserved and combined with the whole. If there are a very large number of elements arrayed on the sensor focal plane (or equivalent), the contributions from fields of detectors may be summed or contrasted. This is what occurs in the human retina, where the number of photodetectors vastly exceeds the number of available fibers in the optic nerve.

Perhaps the complex sensor data of most interest is a spatial array of intensities. Such a synthesis will be multivalent in space, but cumulative within each sensor element. The information contained in a spatial array of intensities is discussed in section 3.3.1.

3.2.5 SYNTHESIS OF BANDS

Of the three kinds of multielement synthesis, that of bands is probably the most difficult. For the sensor bands, there is a spectrum rather than a spatial array. Possible bands -- unlike intensity range or spatial field of view -- are mode-dependent, because every physical interaction has different state parameters. The goal of a multielement synthesis in band is to cover the entire spectrum of interest with a certain detail and resolution. Since the order of the integration depends on the parameters, it is best to consider the band synthesis for each mode separately.

Propagating waves have a frequency, and a corresponding wavelength equal to the wave speed divided by the frequency. The frequency is a positive real number, ranging from zero to infinity. Sensors that detect propagating waves are usually interested in the details of the frequency spectrum. Since frequencies are one-dimensional, so also is the arrangement of sensor elements to cover the band of interest. Human vision covers the light spectrum with three bands of photoreceptors, whose response is perceived as the primary colors of red, green, and blue (Figure 3.2). The responses of the three kinds of cones overlap to cover the whole spectrum continuously. (Transformation of this trichromatic response into a cyclic order is explained below in section 3.3.3.)

Vision is primarily spatial and only secondarily frequency-sensitive. The reverse is true of hearing. The acuity and range of hearing in frequency band is much greater than that of vision. The human ear responds to sound waves from approximately 20 Hz to 20 kHz, and it divides this broad spectrum up into a large number of increments. The hair cells of the inner ear are tuned to different sound frequencies. Hearing, in other words, has not three, but hundreds, of primary "colors". Correspondingly, we are very sensitive to the subtleties of the frequency structure of sound waves. Visual art consists of patterns in space, while music consists of patterns of sound frequencies in time. Frequency information can be translated into spatial extension and then detected as a spatial pattern. In an optical spectrometer, for example, a light ray passes through a prism and is split up into a spectrum of colors arranged linearly by ascending frequency.

Electromagnetic waves, such as visible light and radio waves, have a polarization in addition to frequency. Polarization information has typically been

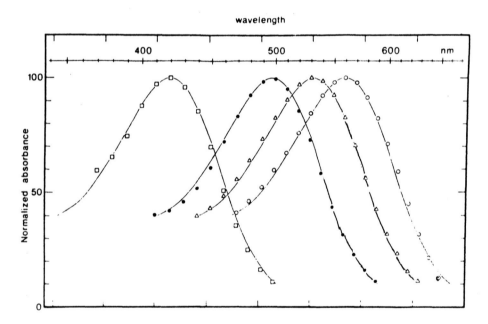

Figure 3-2. Response of Human Photoreceptors as a Function of Wavelength .

squares: blue cones
filled circles: rods
triangles: green cones
open circles: red cones
(Adapted from Reference 16.)

ignored by most sensor systems, whether artificial or biological, because most of the light we receive under normal conditions is unpolarized (e.g., an equal combination of all polarizations).

For propagating objects, the bands become both more complicated and less clear. All propagating objects will have an energy or velocity, and a sensor can view the energy spectrum over a certain band with a certain detail. This, like the frequency/wavelength spectrum of waves, will have a linear order. All propagating objects also have a mass, and this can be measured, for example, with a mass spectrometer. Yet few biological sense organs are directly sensitive to such spectra. Much more common is the distinction of chemical substances, as in smell and taste. The spectrum of chemical substances and their properties does not form a neat linear or circular pattern. Correspondingly, our perception of different smells and tastes falls into broad categories which seem to have little order in relation to one another. We can perceive different scents, for example, but they do not form an orderly sequence like colors or tones.

Force fields seem to have the least potential to be split into bands. Touch, for example, mainly conveys a sensation of intensity, although it does have qualities of vibration or tickling (heat and cold belong to propagating objects on the microscopic level). Likewise, our sense of balance provides us with spatial information, the orientation of the body with respect to the gravitational field and other accelerations.

3.3 THE IMAGE AS AN ARRAY OF INTENSITIES AND QUALITIES

The data representation yielded by a complex sensor is a spatial array corresponding to the sensor's field of view and resolution. Such spatial arrays of information will be termed *images*, regardless of the sensor mode from which they came. Each spatial point in the image array (corresponding to a sensor element) will in like fashion be termed a *pixel*. The image is an array of pixels, just as the complex sensor is a spatial array of sensor elements.

Each pixel in the image is characterized by an intensity (I) and a set of parameters or qualities. If the spatial field is in three dimensions, this image has the form of $A(x,y,z)$. If there are discrete spatial elements, the field of intensities becomes a matrix $A(i,j,k)$. The value at each location in the image is thus:

$$A(x,y,z) = [\ I, q_1, q_2, \ldots, q_n\ ]$$

The q_i are the quantified parameters associated with the bands or qualities. Of all these attributes, intensity is most fundamental. Intensity measures the magnitude of the sensation; as such, it must be present in all sensation. It symbolizes an energy transfer. Qualities, on the other hand, specify *what kind* of intensity is present: they are specifiers or qualifiers of the intensity. Complex sensor images are thus spatial arrays of intensities and qualities. The information in each pixel will be contained in a vector of numbers or symbols corresponding to the intensities and qualities.

Thus, we must attempt to understand in detail exactly what intensities and qualities correspond to. The previous chapter discussed their connection to sensory transduction; it is now time to pick up that discussion where it left off. Table 3-1 briefly summarizes the distinctions and salient characteristics of intensity and quality.

3.3.1 INTENSITY

Intensity is, to repeat, the most fundamental characteristic of sensation, measuring the quantity of energy involved. Intensity is found in all sensation: one can have intensity without any qualities, but not the reverse. It is also the most readily quantized descriptor of sensation, with units coming from the method of sensing. Intensity for a pixel will invariably be represented by a single, positive real number. Intensity is mode-dependent in that its units depend on the mode of sensing, and what appears in one sensory mode may be invisible in another. We use different terms to describe sensible intensities for each mode in everyday life: brightness for vision, loudness for sound, odoriferous vs. odorless for smell, spicy vs. bland for taste, and so on.

The translation of stimulus energy intensity into an internal numerical magnitude need not, to repeat, be linear. In fact, the relation of the two is often logarithmic or geometric. This is certainly true of human sensory transduction, which obeys either the logarithmic Weber-Fechner law or the Stevens power law[4]:

$F = k \log (S/S_o)$ Weber-Fechner Law
$F = k (S - S_o)^n$ Stevens Law

Table 3-1. Intensity and Qualities: A Summary

1.	*Intensity*	*Examples: brightness, loudness*	
	a.	symbolizes total energy or number flux	
	b.	its units come from the sensing method	
	c.	one can have intensity without any qualities	
	d.	is the quantity or magnitude of sensation	
2.	*Simple Qualities*	*Examples: color, pitch, scent, flavor*	
	a.	unextended in space or time	
	b.	specifies the intensity: there are no qualities without intensity.	
	c.	mode-dependent: a translation of the stimulus state parameters.	
3.	*Complex Qualities*	*Examples: shape, texture*	
	a.	extended in space and/or time -- essentially a spatio-temporal pattern	
	b.	distinction of perceived extension:	
		i. microcomplex --	extension is not perceived
		ii. macrocomplex --	extension is perceived
	c.	distinction of pattern basis:	
		i. intensity patterns --	no qualities involved, hence independent of mode
		ii. quality patterns --	depends on patterns of simple qualities like color

where S is the stimulus intensity, S_o is the intensity threshold, k and n are sense-dependent constants, and F is the sensor response as measured by frequency of nerve impulses. The exponent *n* in Stevens Law varies from a low of 0.5 for visual brightness up to a high of 3.5 for electric shock[5]. The average value of *n* for the five classical human senses is about 0.9.

Artificial sensors are also quite various in how they translate stimulus intensity into numerical magnitudes. For an infrared sensor, for example, the response is linear from the threshold to a region where saturation non-linearity begins:

$$Q = ,N + d$$

where Q is the charge produced during the integration (dwell) time, N is the incident photon flux (photons/pixel/dwell), , is the detector efficiency, and d is the dark current charge (an effective threshold). Values of the efficiency , range from 0.02 to 0.5. In the non-linear region, Q approaches the saturation value exponentially, as in the charging of a capacitor. Figure 3-3 illustrates this response function[6].

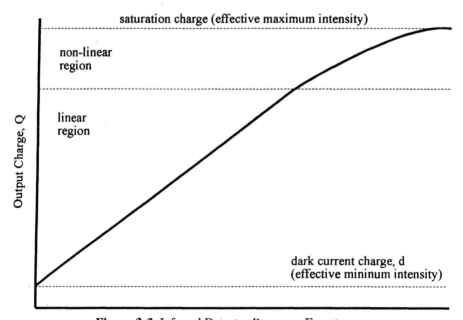

Figure 3-3. Infrared Detector Response Function.

To return to an earlier discussion, the complex sensor data of most interest is often a spatial array of intensities. The data representation will be an array of binary numbers, which represent the intensities at each point. If there are a total of n sensor elements and each element has m_j intensity levels, then the maximum information yield of the sensor array, in bits, will be:

$$i = \sum_{j=1}^{n} \log_2(m_j) \quad \text{bits.}$$

If all the m_j are equal (which is usually the case), then the information yield will be:

$$i = n \log_2(m) \quad \text{bits.}$$

Alternatively, consider the additive synthesis of n elements with m_j intensity levels each. The information contained in the result is:

$$i = \log_2\left(\sum_{j=1}^{n} m_j\right) \quad \text{bits}$$

If all the m_j are equal, this becomes:

$$i = \log_2(m) + \log_2(n) \text{ bits.}$$

3.3.2 QUALITIES IN GENERAL

We have seen that the information output of a complex sensor consists of a spatial array or field, characterized at each point by an intensity and a set of parameters or qualities. It is now time to consider these qualities in more detail. Qualities, to repeat, specify what *kind* of intensity is present at each point. Qualities depend on the mode, and are, in essence, sub-modalities. They are perceived as colors, pitches, scents, tastes, and so on. Patterns in space or time are also qualities.

Quality throughout this work is taken to mean *sense*-quality. The notion of quality originally meant anything that could be attributed to an object. Aristotle, for instance, included habits, dispositions, and capabilities among the qualities[7]. Two kinds of classical qualities are more relevant here: the affective qualities and the geometric properties of objects like shape -- or the proper sensibles and common sensibles[8]. The proper sensibles (e.g., color) could be apprehended by one sense only. The common sensibles (e.g., shape, extent, motion) could be apprehended by each of the senses. This difference provided the distinction for the later division made between primary and secondary qualities. The primary qualities, corresponding more or less to the old common sensibles, were mathematical: extension, figure, location, motion. These belonged to the object itself. The secondary qualities were mode-dependent, specific to each sense, like color. These do not belong to the object, but to the sense organ itself.

Chapter 9 will take up the question of the correspondence of qualities to the nature of objects. This section's aim is to introduce a new division of qualities, that of *simple* and *complex*.

A *simple quality* has no extension in space or time or band. It is essentially "pointlike". It can be perceived by a single sensor element. Examples of simple qualities are colors, pure musical tones, single smells, and so on. Musical harmonies are also simple qualities, as will be explained.

A *complex quality*, in contrast, is a certain kind of pattern in space or time. Complex qualities have some extension and require multiple sensor elements to detect. They depend on a field of relative intensities -- of two or more intensities perceived together. When we see textures, or distinguish between shiny and dull surfaces, we are perceiving complex qualities: these are spatial patterns. (An unextended object cannot have complex qualities. A star, for example, has color and brightness, but not texture or dullness.)

Simple qualities, as the name implies, can potentially be the output of a simple sensor or protosensor. Complex qualities can only be grasped through the use of a complex sensor. Only a complex sensor can provide the extended field of multiple intensities on which complex qualities depend. They require a multiplicity perceived as a unity.

3.3.3 SIMPLE QUALITIES AND THEIR REPRESENTATION

The simple qualities are perceived representations of the state parameters of the physical stimulus. Here we consider just how these parameters are translated by the sensor. For each point in a spatial field or array, there is an intensity I and several parameters ($q_1, q_2, \ldots, q_n$) describing the qualities. Color vision, for example, involves three intensities corresponding to the overlapping wavelength bands of the three kinds of cones in the retina. These intensities are synthesized and translated into perceived colors:

700 nm: red intensity (R)	brightness (I)
546 nm: green intensity (G) →	hue (q_1)
435 nm: blue intensity (B)	saturation (q_2)

The trichromatic intensities are converted into a single, unified intensity (brightness), plus the two qualities of hue and saturation. There are several different ways of parameterizing color. A simple and straightforward one is a color circle with equally-spaced primary colors of red, green, and blue:

$I = R + G + B$	total intensity
$q_1 = \arctan(y/x)$	hue (an angular measure)
$q_2 = (x^2 + y^2)^2 / I$	saturation (a radial measure)

where

$$x = R - \frac{1}{2}(G + B)$$

$$y = \frac{\sqrt{3}}{2}(G - B)$$

Although light frequencies are a linear series, the synthesis of the three chromatic values takes a circular form: red and violet, although at opposite ends of the frequency spectrum, visually resemble each other. This leads to the color circle, first discovered by Isaac Newton. The quality parameters of hue and saturation define the place of any color on the color circle. The hue is the angle and the saturation is the radius. (Figure 3-4). The hue is what we normally term color -- red, yellow, green, blue, and so on. The saturation is the degree of purity of the color -- colors with zero saturation are the neutral shades of gray, black, or white.

The trichromatic intensities are combined in accordance with a rule that yields a single intensity and the cyclically ordered hue and saturation. Each actually perceived specific color may be found in this generic "color space". We are aware that individual colors belong to the same genus of "color", even though we never perceive "color" itself, only individual colors. Each sensory mode defines a genus or set of specific perceived qualities, and the qualities themselves are some mapping of the parameters of the physical interaction. The simple qualities thus nearly always correspond to the sensor band. (As a practical concern, there is a difference between the primaries and mixing of colored pigments and those of colored light. Pigments are subtractive: a yellow pigment reflects those wavelengths of light perceived as yellow and absorbs all the others. Thus a mixture of pigments will subtract more of the visual spectrum to yield a darker color; a mixture of all pigments would yield a dark gray or black. Colored light, however, is additive: a mixture of all hues of light yields white.)

Simple qualities may thus involve a combination of several different band stimuli: different wavelengths of light combined are perceived as a single color. Such simple qualities are *structured*. A good example of structured simple qualities are musical harmonies. They are not extended in space or time, but they do contain several pure tones that are perceived together in one distinctive sound. The same is true of the spectrum of harmonic tones in a musical instrument or voice.

Qualities in an information system are necessarily quantized (actually or effectively reduced from a continuity to some finite-length numerical representation), but the numbers that stand for each quality are of an entirely symbolic nature. The qualitative meaning is retained in this way. The number that stands for a certain frequency of light does not mean the same thing as a number that stands for the detection of a particular chemical substance, and they may not be directly combined. They belong to different sensory modes, have different units, and thus entirely different meanings.

There is good evidence for the preservation of qualitative differences in the human nervous system. On the one hand, all stimulus intensities are represented

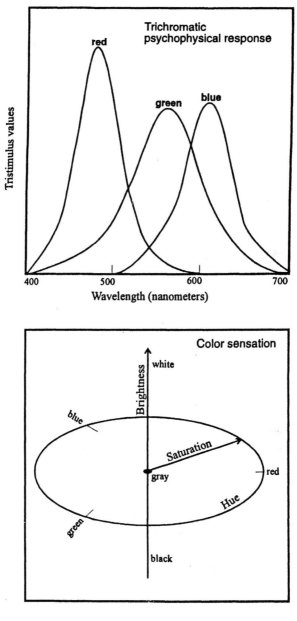

Figure 3-4. Translation of Light into Color Hue and Saturation[9]

by the frequency of nerve impulses. On the other, the distinctions of sensory qualities is carefully maintained. Intensities of different qualitative kinds travel along different nerve fibers. For example, different nerve fibers in the auditory nerve are tuned to different sound frequencies[10]. In the optic nerve, there are nerve fibers which deal with only the complementary colors of red and green, and others which represent only yellow and blue. This accounts for the constancy with which different people and different nations identify colors and their order[11].

"Rosch (1977) has convincingly shown an impressive degree of uniformity in color perception across a wide variety of different cultures. ... The differences have to do with where people in different cultures locate boundaries between colors. Thus, the cross-cultural evidence, when gathered and interpreted properly, supports the conclusion from developmental studies that basic color experience derives from a built-in, essentially fixed neural coding system."

In other words, there is an identical representation scheme for colors, regardless of subjective variations in the sensitivity of the retinal cones, cultural influences, and so on. Defects in color vision, like color blindness, do not change this fact. In color blindness, part of the visual system (usually the red-green system) is dysfunctional, but the order of the colors is unchanged.

Qualities need not be ordered in a circular fashion as colors are. Auditory pitches are arranged in an essentially helical way, the symmetry axis of the helix being frequency and the radial direction being the position in each octave. The sensor form is essentially a transformation rule that describes the relation of the qualities to each other. For some qualities, such as smell, there is very little order among the different perceived qualities. (It is worth noting that such senses are spatially vague. Conversely, the complete synthesis of color information is necessary for the highly developed spatial aspects of vision.)

Consider a hypothetical creature whose vision is sensitive not to wavelengths of light, but instead to its polarization. Polarization of light is a complex phenomenon which can be explained here only schematically. Individual photons, the quanta of electromagnetic radiation, are either right-circularly polarized (RCP) or left-circularly polarized (LCP). But the light we sense under ordinary conditions consists of vast numbers of incoherent photons. This means that most light we see is unpolarized. But if the light is coherent, or has been passed through a polarization filter, then it can be polarized. These

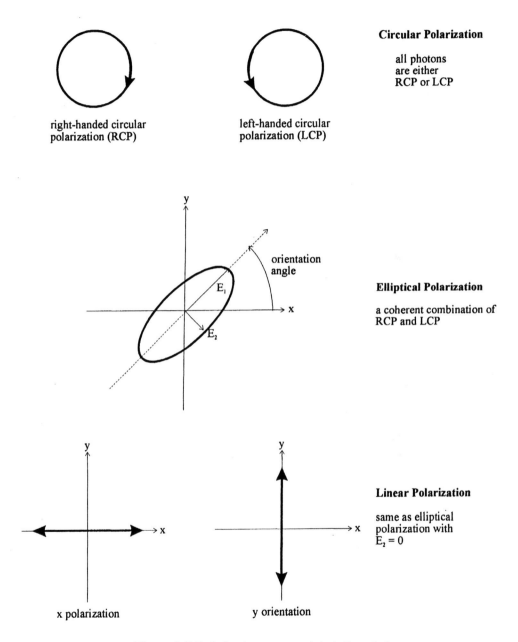

Figure 3-5. Polarization states and their Translation.

polarization states result from a combination of LCP and RCP with a phase difference between the two. Linearly-polarized and elliptically-polarized light are obtained in this way (Figure 3-5).

Polarization states can be described in a number of ways. Here, we will use four parameters: (1) the major axis of the ellipse, E_1, (2) the minor axis of the ellipse, E_2, (3) the orientation angle of the ellipse, varying from 0 to 180 degrees, and (4) the net helicity: +1 for right-handed and -1 for left-handed. Circularly polarized light will have $E_1 = E_2$, while linearly polarized light will have $E_2 = 0$.

Let us suppose that our hypothetical creature's eye is sensitive only to the total energy intensity (brightness) and the orientation angle at each point in the retinal image. It will ignore the helicity, but be sensitive to the degree of linear polarization. (This means that RCP or LCP light -- which is not oriented -- will be perceived the same as unpolarized light.) The polarization intensities can then be parameterized as:

$$\text{Intensity} = E_1 + E_2$$
$$q_1 = 2 * \text{orientation angle}$$
$$q_2 = (E_1 - E_2) / (E_1 + E_2)$$

Further suppose that the parameters q_1 and q_2 were perceived as colors: the first playing the role of hue and the second of color saturation. Pure hues would correspond to linearly polarized light. The neutral tones of white, gray, or black would correspond to unoriented light (either incoherent or circularly-polarized). Tints, tones, and shades like pink or brown would represent elliptically-polarized light or a mixture of linearly polarized light with incoherent light.

Due to their dependence on both mode and the sensor itself, kinds and orders of simple qualities and their data representations are various. Nonetheless, all simple qualities must possess certain common features. There will always be a single intensity associated with the qualities, which is qualified or specified by them. There will also be a neutral combination of qualities like the colors gray, white (maximum intensity), and black (zero intensity). This is why sound that consists of an incoherent mixture of all frequencies is called "white noise". Correspondingly, there will also be a parameter which describes the purity of the quality, such as the saturation of color. Every simple quality may thus be represented as:

[intensity, saturation, one or more quality parameters (e.g., hue)]
A two dimensional color image A(x,y) would be represented as:

A(x,y) = [intensity, saturation, hue]

For a black and white (gray scale) image, however, there are no qualities other than intensity I and A(x,y) = I(x,y).

The information content of a complex sensor image can be obtained by expanding upon the approach described in section 3.3.1 for intensity arrays. Qualities or band parameters will generally be combined in a multivalent fashion. If each quality q_i has n_i increments, and the intensity has m levels, each point in the spatial array will contain an information of:

$i = \log_2(m) + G \log_2(n_i)$ bits.

Suppose, for example, that we have a sensor element for which $m = n_1 = n_2 = 256$. This implies an information content of i = 24 bits per element. The information content of a VGA array of 640 x 480 = 307,200 such elements is 7.37×10^6 bits (0.92 Megabyte). Thus, the old line that "a picture is worth a thousand words" actually falls short of the mark -- most images contain information far exceeding a thousand binary words. The vast amount of storage space that image data (and, increasingly, video data) requires has been a driving force behind the steady growth in the sizes of computer memories and disk drives.

3.3.4 COMPLEX QUALITIES

Complex qualities depend upon the comparison and contrast of two or more sensor elements. They are all patterns requiring a spatio-temporal structure. As was introduced in Table 3-1, the complex qualities are subdivided into two kinds depending on how the patterns are perceived: microscopic and macroscopic.

For microscopic complex qualities (*microcomplex* qualities, for short), the extent of the pattern in space or time is not directly perceived. It is perceived as a coherent whole, without knowing its microscopic structure. An example in vision is the distinction between shiny and dull surfaces. Such a distinction depends on relative intensities on the microscopic level -- shiny surfaces are smooth, and dull surfaces are rough. But we do not see this microscopic structure; what we see are shiny and dull without knowing its cause. Generally, microcomplex qualities are based on repetitious or quasi-random (self-similar) patterns. Textures are microcomplex qualities to both vision and touch. Hearing is particularly rich in microcomplex qualities. They include the distinctive sounds of different instruments, different tones of voice, different phonemes. When we hear a spoken syllable, what we hear is the syllable and its phonemes, and not the details of the waveforms.

The microcomplex quality represents a symbolization of spatial or temporal structure. Just what will be perceived as a microcomplex quality depends on the abilities and limitations of the information system and sensor involved. Human vision has several channels tuned to various spatial frequencies, in order to detect repetitious patterns[12]. This is apparently done to preserve texture information while not overloading the visual system with too many details. Thus, we see many surfaces that seem smooth, yet structured, like velvet, the coat of a cat, or a wheat field[13]. We do not consciously perceive surface corrugations if their spatial frequency is higher than about 4 cycles per degree. Such patterns are perceived instead as microcomplex qualities. Microcomplex qualities are also responsible for many of the perceptual coherence laws of Gestalt psychology, and they play a major role in the visual identification of objects (different objects have different textures and shading).

There is one epistemologically interesting point to be drawn from the microcomplex qualities. The existence of such extended qualities, grasped as wholes, means that we necessarily grasp spatial and temporal relations in sensation. Without such relations, complex qualities of any sort could never be perceived.

For macroscopic complex qualities (*macrocomplex* qualities), the spatio-temporal extent of the object and its patterns are obvious. Indeed, it is often the information sought. Macrocomplex qualities thus correspond to the old "primary qualities" of extension, figure, motion, etc. It is from them that we extract spatial and temporal information about the world. The perception of a melody or an entire word are examples of macrocomplex qualities in hearing. The color harmony of a visual pattern is a macrocomplex quality.

Complex qualities, macrocomplex qualities in particular, depend upon a spatial or temporal field of *varying* intensities. Our visual grasp of objects, for example, depends upon being able to detect edges (sharp intensity variations), as well as more subtle variations of shading. Already in the retina, the edges of objects are enhanced through the effective second derivative of the spatial field[14]. The *features* of objects, depending once more on a spatial variation of intensities, are very carefully arranged in the brain. Without intensity variations, the grasp of the so-called primary qualities would not be possible. Conversely, the fact of our perception of such qualities proves that we can grasp an extended field of intensities as a single whole. We perceive shapes as wholes and not as disconnected congeries of parts.

Complex qualities based on intensity patterns are not essentially dependent on the sensor modality, as all sensor modes have some spatio-temporality. There can be, however, mode-dependent complex qualities. These are spatial or temporal patterns of simple qualities. A melody, for example, is a mode-dependent complex quality. Vision is sensitive to many different color patterns, harmonies, and contrasts. A full consideration of mode-dependent complex qualities would take us deep into the theories of art and music. We must note something of immense importance to aesthetics here: sensate beauty is essentially a valuation, a classification, of complex qualities. That must be so because beauty always depends on a proper relation of the parts to the whole -- on a pattern of some sort. It is thus based on a complex quality pertaining to an entire image.

In a spatial or temporal array of intensities and qualities, there are as many potential complex qualities as there are patterns. This means that for an array containing n bits of information, there are 2^n potential patterns. This also means that the data representation of complex qualities can often be more difficult than for simple qualities. Microcomplex qualities usually depend on a repetitive structure and can be effectively represented by a small set of parameters. Random patterns can be represented by a phase spectral density, while various shapes can geometrically described. Or, distinctive patterns can simply be represented by discrete symbols, as phonemes are. A full consideration of macrocomplex qualities begins to overlap with the field of pattern recognition, which is overviewed in Chapter 10.

3.4 EXTENSION AND INTENSION OF COMPLEX SENSORS

One of the main purposes, to repeat, of combining simple sensor elements to form a complex sensor is to cover a wider extent -- in intensity range, in field of view, or in band -- yet preserve the finer detail of the simple sensors. Complex sensors, like simple sensors, have an extent in each of these dimensions, but they also have a new aspect that simple sensors do not have: a series of levels or increments or regions within the extent. This is the *intension* or *resolution* of the sensor.

Since intension can be confused with intensity, the term resolution will generally be preferred. Intension, however, indicates its complementary role to extension. The more extension a sensor has, whether in intensity, field, or band, the more objects it can apprehend. It can "see" further and wider. The more intension a sensor has, the more detail it can convey -- it provides more information about each object. It measures the acuity of sense.

Every actual complex sensor will have a trade-off between extension and resolution. An sensor with great extension is like a net of immense size, but equally open mesh, cast over the world. The information it provides will be broad, but vague. It will be general. The converse case is a sensor with fine resolution, but small extent. That kind of sensor is like a microscope, aimed at one small area. The information yielded by an intensive sensor will be narrow, but very detailed. High resolution means specificity or specialization. The protosensor has infinite resolution, but infinitesimal extension.

For a spatial or quasi-spatial grid of sensor elements, extension and intension can be Fourier transformed into each other. This indicates once more their essential complementarity. If we have a spatial image array $A(x,y,z)$, there is a corresponding spatial frequency array:

$$B(k_x,k_y,k_z) = F(A(x,y,z))$$

where k_x, k_y, k_z are the wavenumbers in the x, y, and z dimensions, and F is the Fourier transform. The Fourier transform in one dimension is:

$$B(k) = F(A(x)) = \int_{-\infty}^{\infty} A(x) e^{-ikx} dx$$

In the B array, the lower spatial frequencies will contain the gross features of objects, like image contrast. The higher spatial frequencies (the extremities of the B array) contain sharp edges and fine detail. Removal of this part of the array, and transformation back to ordinary space will give an image A' that is blurred and less detailed. Thus, a wider B array means not more extension, but more intension. The spatial image A contains the objects ordered as they actually appear, but the spatial frequency array B contains the information content of the objects ordered by type of pattern. Entire patterns can be removed from an image by transforming A into B, blocking out a specific portion of B, and then retransforming to yield an new image A'. This property is taken advantage of in optical spatial filtering and holography[15]. Conversely, selecting a certain portion of the transformed array B means seizing upon a general aspect of all the objects in the image. Chapter 10 discusses the image processing applications of the Fourier transform further.

Extension and resolution define, in essence, the phase space of the sensory image. Phase space originated in physics to describe the states available to a system. It consists of corresponding pairs of spatial coordinates (x,y,z) and momenta (p_x,p_y,p_z). The phase space of a spatial image will consist of the spatial coordinates (x,y,z) and the spatial frequencies (k_x,k_y,k_z). Any object sensed will have a location and extent in phase space. In the spatial coordinates, this is just how it appears to us. But its size or volume in frequency space indicate its complexity.

CHAPTER FOUR

MULTISENSOR FUSION

4.1 GENERAL ASPECTS OF MULTISENSOR FUSION

In *multisensor fusion*, two or more sensors of different modes or methods are united to form a single sensor system -- a multimode sensor or sensor suite. Multisensor fusion also deals with the problem of combining two or more widely separated complex sensors of the same mode, as in binocular vision.

The central problem in multisensor fusion is to unite sensor data from different modes into a single representation or image. The quantitative study of sensor fusion was developed in the practical context of artificial sensors, and it is so new that it is a field without well-defined techniques and terminology[1]. The multisensor fusion problem is inherently more difficult than construction of complex sensors, for the outputs of the various sensors each have a different meaning. Sensor fusion requires some commonality among different modes of sensors to integrate their data into a one image. Presumably, such data fusion is intrinsically similar for both artificial sensor systems and biological sense organs.

Despite the difficulties involved in multisensor fusion, there are such strong advantages to having several sensor types that it has been extensively studied[2,3]. Integration of different sensor modes supplies a much more *complete* picture of the sensor's environs than a sensor of a single mode, regardless how sophisticated. That is why nearly all animals possess several different modes of sensation. Vision, hearing, touch, smell, and taste all "view" a different aspect of the world, and each conveys information inaccessible to other modes. Each sensor mode can provide *different kinds* of information appropriate under different circumstances, or in conditions where other sensors cannot operate. For example, touch provides us with spatial information even in the dark, where vision cannot, while vision provides the same information about distant illuminated objects that are out of reach. Use of multiple sensor modes allows one to correct the uncertainties of each sensor; as there is mutual confirmation of the presence and properties of objects. They can provide resistance to "jamming" and camouflage, as these tend to be mode-dependent. Multiple sensor systems also have an inherent redundancy, while a single sensor system would be crippled by the loss or malfunction of that sensor.

In short, multisensor data fusion leads to more complete images of the world, which in turn leads to more accurate inferences -- and different kinds of inferences -- from sensor data. For a living being, this means an increased chance of survival. It also means an expanded set of capabilities -- an expanded world, so to speak. For an artificial sensor system, it means a similar increased probability of success.

The growth of the study of multisensor fusion has been driven by several practical applications[4]. Probably the most extensive application has been to military intelligence and weapons systems. There is an increasing need to integrate the data from many surveillance sensor systems (radar, optical sensors, etc.) into a single picture of the battlefield or the disposition of enemy forces. At the same time, most aircraft, ships, and tanks will possess several sensor systems, whose data must be fused for the use of their commanders. Remote sensing -- the use of several sensors over a wide area to survey geological or agricultural conditions -- is another practical application of multisensor fusion. Medical diagnosis using multiple sensors (e.g., magnetic resonance and patient biological data) is a growing area of application. Perhaps the ultimate application of data fusion for multiple artificial sensors is in robotics. Many of the above applications are in their infancy and will expand greatly with improved data fusion algorithms and computer systems.

4.2 GROUNDS FOR MULTISENSOR DATA FUSION

To repeat, the synthesis of sensor data from different modes into one image requires some aspect or property common to all the sensors. Without a commonality, it is impossible to attribute different properties to the same object. To use a classic example, we perceive that a sugar cube is both sweet and white. Two different sensor modes (taste and vision) are involved here. There must be attributes of the sugar cube, regardless how vague, that are present to *both* taste and vision to make this unified perception possible. Conversely, when there is no commonality, no objectively reliable multimode data synthesis is possible -- combinations of data from different kinds of sensors would be at best a happy accident. The crucial point is thus to find what grounds of multimode data fusion are possible.

The commonality cannot be of band or simple quality, for these are mode-dependent. Likewise, it cannot be the intensity, because the meaning of intensity is different for each sensor mode. To try to unite multimode sensor data on the basis of simple quality or intensity would be to jumble up apples and oranges. Fusing data on the basis of simple quality is to compare and contrast *different* objects, rather than to unite the attributes of the same object. To unite different objects by their qualitative likeness is not multisensor fusion, but a rudimentary kind of concept formation or pattern recognition, which we will return to in Chapter 10.

The commonality that permits multimode data synthesis must be present for each sensor mode employed, yet not dependent on the mode as such. It must allow reliable discrimination of different objects. This leaves three possible grounds of multisensor fusion. Any of the three will allow the various modal qualities to be assigned to their respective objects:

1) The first is *extrinsic* or *spatio-temporal fusion* -- the commonality is a shared position in space, point in time, or kind of movement. Such properties will be present for all sensor modes: all sensors have a field of view. Spatio-temporal fusion is also called quantitative fusion, from its amenability to mathematical techniques.
2) The second possible kind of multisensor fusion is on the basis of

something intrinsic to the objects themselves -- an *intrinsic fusion*. This depends on a complex quality, because complex qualities, as spatial or temporal patterns, can be present in several sensor modes.

3) Finally, there is the case where the identities of the objects are known to each sensor mode. This is *direct fusion*. Direct fusion applies if there is just a single object present: whatever is sensed, by default, must come from that object. Another situation well-suited for direct fusion is when the objects themselves actually announce their identities, as with aircraft transponders. (Direct fusion also applies when dealing with specific objects on a network, rather than objects in physical space. See Chapter 11 for a consideration of this problem.)

Multisensor data fusion also depends on the nature of the information sought. The goals of the larger system to which the sensor belongs (a human being, a jet aircraft, etc.) apply explicitly. In a *positional fusion*, the primary goal is to obtain spatial information regarding the objects -- to answer the question "where is it?" more accurately. Spatial resolution and/or extent is improved through the use of several sensors[5,6]. The resulting image array from a positional fusion will have more "grid points" than any of the images that compose it. In *identity fusion*, by way of contrast, the goal is to determine the identity or properties of the objects in the image. It seeks a better answer to the question "what is it?" The information from different modes will be kept separate. These two kinds of goals, positional or identity, limit the sensor integration method that can be used. Identity, for example, will require an extrinsic or spatio-temporal fusion of data.

4.3 SPATIO-TEMPORAL FUSION

In spatio-temporal or extrinsic fusion, the commonality that permits the synthesis of different sensory modes is quantitative, mathematical, and geometric. It is based on space (collocation), time (simultaneity), motion, or shape. Spatio-temporal fusion is possible for two reasons. First, all sensory objects exist at particular points or regions of space and time. Second, all sensors have some spatiality, regardless of how vague. Each views some region of space and time. This means that if sensors of different modes are "lined up" properly, viewing the same region of space (or if the location of the center of the image is known), then qualities from different modes can be correctly assigned to their appropriate objects.

The geometric commonalities that permit spatio-temporal fusion are none other than the complex qualities. They are the macrocomplex qualities of an entire image, rather than properties of individual objects in the image. Complex qualities, being spatial or temporal patterns, are essentially mode-independent and can appear the same to each of the senses. This concurs with the older notion of "primary qualities" or "common sensibles" that could be grasped by all the senses[7].

Spatio-temporal fusion requires knowledge of the sensors' locations in both space and time. For a human being, they are all "here", while the objects are all "there". Many artificial sensor systems, however, have sensors widely distributed in space -- a network of radar for air traffic control, for example. Even when sensors are on the same platform (e.g., a satellite), one must know the spatial location and orientation of the sensors before the data synthesis can be performed.

The more accurate the spatial information available from each sensor, the more accurate the quantitative sensor fusion can be. This can be a problem if one attempts to unite the data from two sensors of widely varying spatial resolutions. If I simultaneously see and hear a bird in a tree, my assignment of the bird's call to the visible bird will only be as accurate as the spatial acuity of my hearing. If a single bird is present, that is unimportant; but if several different kinds of birds are perched in the same tree, it matters a great deal. Generally, spatio-temporal fusion requires sensors of comparable resolutions.

A primary advantage of spatio-temporal fusion is its quantitative nature, allowing the use of powerful mathematical and statistical techniques. Objects are described with respect to each other according to the laws of geometry and physics. One of the most common quantitative data fusion techniques is Kalman filtering, or some variant of it. The Kalman filter is a linear decision rule, which recursively calculates an estimate of a parameter vector on the basis of previous estimates and new observations[8]. Kalman techniques are particularly useful when many moving objects are being tracked. The goal of all such positional fusion techniques is to obtain an estimate of the state vector (position and velocity) of the objects. Once this is known, data synthesis becomes possible.

There are two possible approaches to spatio-temporal fusion. The first is a direct overlay of collinear spatial images. This requires that the sensors being used are collocated on the same platform, such as radar on a ship or eyes in one's head. The sensors all have to be at the same place and looking at the same "target". The second -- the norm in remote or distributed sensing -- is to determine the location of the objects for each sensor image first. Then, the various sensor images are joined or fused together according to their respective objects or regions of space they cover.

Although sensors of different modes are usually separate, this need not be the case. Multimode sensor fusion on the "microscopic" level, where sensor elements of one mode are intermixed with those of another, is entirely possible. We can imagine, for example, an array of sensor elements in which photodetectors alternate with radiation particle detectors. The spatial sensor fusion here is obvious, because the two kinds of sensor elements are on the same spatial grid. This sensor would provide an image of the object that is both optical and "radiative". (Conversely, the radiation detectors could be used to "cancel out" radiation noise in photodetectors.)

Unification can also be on the basis of motion -- objects will have distinct velocities no less than locations. We both see and hear a fire truck as it passes by, for example. The movement involved can be oscillatory, such as hearing and touching a ringing bell.

4.4 INTRINSIC FUSION

The key to intrinsic fusion is the ability to recognize the objects for every sensor mode being used. Intrinsic fusion thus depends on a partial pattern recognition and the cooperation of the information system. (In practical use, intrinsic fusion generally applies some prior knowledge of the properties of the objects involved.) If the pattern is present for each sensor mode, it can only be something spatial or temporal, but intrinsic to the objects themselves. These can only be complex qualities: unlike simple qualities, they are extended in space and/or time. Whether by macrocomplex or microcomplex qualities, an intrinsic fusion further requires that each object have a unique pattern, its own individual "fingerprint" or "signature".

Macrocomplex qualities include the shape of an object, its particular extent, and its macroscopic internal structure -- the image of its parts. Here the data fusion will be accomplished by matching shapes, wave envelopes, and so on. Intrinsic sensor fusion can also proceed through microcomplex qualities. Both macrocomplex and microcomplex qualities are spatio-temporal patterns. Thus, sensor fusion could conceivably take place on the basis of texture -- simultaneously feeling and seeing a textured object, for example. In both cases, the texture means the same thing -- a spatial pattern -- and such a commonality will permit joining the two modal sources of data.

There is nothing to prevent a combination of techniques of intrinsic fusion and spatio-temporal fusion. The use of known pattern information can be used to align sensors of varying spatial acuity.

4.5 POSITIONAL FUSION

The goal of positional fusion is to sharpen spatial resolution or increase extent through the use of several sensors. It can also be used to infer a new spatial dimension, as in binocular vision. It is typical that for positional fusion that several identical or very similar sensors of the same mode are used.

Multiple sensors of different modes can also increase the spatial information available. An example is simultaneously seeing and touching an object. The spatial acuity of vision in the directions of height and width is excellent, but, even with binocular vision, its information regarding depth is much less precise. Touch remedies this lack. A very similar improvement of information can be found in artificial sensors in the combination of an infrared optical sensor with a radar[9]. An optical sensor, like the eye, has excellent horizontal and vertical spatial resolution, but little or no depth perception. A radar, in direct contrast to the optical sensor, provides excellent information regarding the range of objects. Combination of the two gives a much more precise estimate of the spatial location of objects.

Let us return for a moment to binocular vision. We will not go into any great detail here, but aim to seek the essential factors that make it possible and the information that is gained by using it. Binocular vision is not a multimode sensor integration, because it involves two optical sensors. But it resembles multimode integration in that it uses two complete and separate complex sensors as a single system. The object of binocular vision is indeed positional: from two separate two-dimensional sensor images, it discerns information of the third dimension, depth. It attempts to fuse two flat images into one three-dimensional one.

For a single point, determination of depth from two separate observations is a simple exercise in trigonometry. But stereo matching becomes much more difficult when there are many objects in a scene[10]. The fusing of the two images into one depends upon matching corresponding points in the two. This can be done if a given point has a unique location at a given time and the scene is separated into objects with generally smooth surfaces. In other words, the shape and texture information of the objects matters, and this in turn depends upon surface patterns, shadows and shading, surface orientations, and so on. Marr identified three conditions of stereo matching between two images as compatibility of the objects in the scenes (same objects present in each), the uniqueness of each object within each scene, and continuity: the disparity of matches (due to different viewpoints) varies smoothly across the image. In other words, stereopsis depends to a very large degree on the matching of complex qualities in the two images involved. Spatial patterns possessed by objects make this possible. From the slight difference of patterns of the two images, the depth information can then be extracted. In theory, there is nothing that prevents simple qualities (colors) from being used in stereo matching, but in fact the mode-independent complex qualities are used -- it is the spatial information that matters.

4.6   BIOLOGICAL SENSORY INTEGRATION

What has been discussed to this point on sensor fusion has concentrated on the integration of artificial sensors, because the techniques are explicitly known. This is not the case in the joining of the senses in an organism to form a single perception of the world. The details of such integration are within the nervous system and are not easily observed or inferred. Nonetheless, despite the differences of our naturally-supplied sensory modes, they provide us with a unified experience of the world, and we constantly use them in concert without any conscious effort[11]. This means sensor fusion takes place in the human nervous system as a part of the normal sensory process. If there were no convergence of the senses, having a multiplicity of senses would be worthless, because they could never be used to coordinate a single action: it is the fusion of multimodal sensory information that makes it practically useful. Indeed, there is no animal with a nervous system which holds sensory representations of different modes completely separate from each other[12]. This convergence of the senses occurs in the superior collicus of the brain.

Whether sensory fusion is innate or learned is a topic of debate among psychologists, but it need not detain us. The important point is, even if the unification of the senses is learned from experience, as Piaget holds, there must some ground that permits the unification to occur. In other words, there must be some commonality among the senses. In human perception, there are mode-independent (amodal) properties that can be transferred across sense modalities. Experiments have shown that if human observers inspect objects with one sensory mode, they can later recognize them through a different mode: the recognition of shapes through vision, then touch, for example. The amodal properties that permit this are intensity, form, number, and duration[13]. In other words, they are spatio-temporal patterns or complex qualities, just as for artificial sensor systems.

There may even be integration of mode-dependent qualities in the brain. Integrated human sensory experience is richer than the sum of its parts: for example, aroma and texture are components of our experience of the taste of food[13]. In the abnormal condition known as synaesthesia, simple qualities of one mode may evoke the perception of simple qualities of another mode: a pure tone may also bring about the perception of a color, for example. This points to the existence of intermodal qualities, but there is not space here to explore this interesting topic further.

4.7 THE MULTIMODE SENSOR IMAGE

Multimode sensor fusion produces a unified image containing information from each of sensory mode. Intensities and qualities from each mode are assigned to particular objects or spatial locations. We saw in the previous chapter that the complex sensor image is a spatial array with a vector of intensity and qualities at each point, $A(x,y,z) = [\ I, q_1, q_2, ..., q_n\ ]$. A *multimode sensor image* is an expansion of this, with each sensor mode introducing a new "dimension" or vector of intensities and qualities. This yields a two-dimensional matrix for each point in the image array:

$$A(x, y, z) = \begin{bmatrix} I_1, q_{11}, q_{12}, ..., q_{1n} \\ I_2, q_{21}, q_{22}, ..., q_{2n} \\ ... \quad ... \\ I_m, q_{m1}, q_{m2}, ..., q_{mn} \end{bmatrix}$$

where m is the number of modes and n is the maximum number of qualities for all the modes. This combination is multivalent, so the information of the unified image is sum of the information contents of the modal sensor images. The information content will be the number of elements in the array multiplied by the information at each point in the array:

$$i = \sum_j (N_j + \sum_k M_{jk}) \quad bits.$$

where N_j is the information measure for each modal intensity I and the M_{jk} are the information for each quality q in the matrix.

We have seen also how it is possible to rearrange the various modal band parameters into qualities -- the combination of trichromatic intensities into the circular field of colors, for example. Is anything similar *objectively* possible between modes -- are there objective intermodal qualities? There cannot be, because the qualities are mode-dependent, being derived from the parameters of the physical carrier. The comparison and use of multimode qualities and intensities belongs to pattern recognition and artificial intelligence. It fall outside the scope of sensor theory as such. The multimode sensor image is thus the end product of the sensory process as such. It is the maximum that can be obtained from a sensor system.

4.8 THE ARGUS SENSOR

The protosensor was the smallest and most rudimentary imaginable sensor. Now, we are ready to describe the opposite limit, the biggest and most complex theoretically possible sensor system. Let us call this ultimate in sensory capabilities the *Argus sensor.*

We can obtain the properties of the Argus sensor by extrapolation. It would be a multimode sensor system, capable of detecting every kind of physical interaction: all varieties of force fields, chemical substances, particles, and waves. In other words, it would cover all possible modes of sensation. It would thus consist of many sensors integrated into a single system.

The sensors that compose the Argus system would all be complex sensors. They would each have an infinite extent in space, as well as an infinite spatial resolution. Each complex sensor would cover the entire parameter space for its mode with an infinite resolution. It would, for example, have an infinite bandwidth in electromagnetic waves, detecting every possible frequency from radio waves through light up to gamma rays with infinite precision. Each element of the complex sensors belonging to Argus would have an infinite range of sensitivities, with an infinite number of intensity levels.

The Argus sensor, like the protosensor, is an idealization. If the protosensor produces a single bit of information, the information yield of the Argus sensor would be infinite, even for a single time.

The Argus sensor, although practically impossible to build, has an importance central to general sensor theory. The Argus, as we have seen, is the theoretical upper limit for sensors, just as the protosensor was the theoretical lower limit. All actual sensors are subsets of the Argus sensor, and all sensor images are corresponding subsets of the Argus image -- just as all actual sensors can be represented as a combination of infinitesimal protosensors.

The Argus sensor would provide the theoretical maximum amount of information about the world accessible to a purely passive sensory system. As we shall see in later chapters, it can view every aspect and every property of all physical objects. That is so because the Argus sensor possesses all possible sensory modes, and hence encompasses all physical interactions. It provides the truly *comprehensive* view. This means that Argus, like the protosensor, has an epistemological importance. The protosensor showed that sensors can be valid; the Argus shows they can provide -- at least in theory -- a complete knowledge of physical objects.

We can imagine a finite approximation to the Argus sensor, similar to the relation between the simple sensor and the protosensor. Such a sensor would have finite extent and resolution in spatial field of view, in band, and in intensity. It represents the *practical* upper limit for sensor systems. A finite Argus sensor could actually be built, and might prove useful on space probes to other planets. A subset of the Argus sensor -- one covering all frequencies of electromagnetic radiation, for example -- could be built and would find many applications.

CHAPTER FIVE

TEMPORAL CONSIDERATIONS AND VARIABLE SENSORS

5.1 DYNAMIC ASPECTS OF SENSATION

Sensors to this point have been described in purely static terms, as if their parameters and the sensory world were unchanging. Even where time has entered indirectly, as with wave frequencies and complex qualities of waves, it has been considered in an essentially timeless and "spatialized" way. Now we must turn to the explicitly dynamic aspects of sensation. These dynamic problems include:

1) the dynamic response of sensor elements.
2) the analysis of the continuous temporal stream of sensor data into discrete "frames" at specified times.
3) the representation of non-temporal information within a temporal series -- trading time for spatial extent, for example, as in scanning.
4) sensors with variable or controlled ranges of response to intensity and other stimulus parameters.

In all the above points, a continuous temporal stream of data is assumed.

Even if the stream consists of discrete events at a few specific times, continuous monitoring of the stream is still required (unless there is prior knowledge of the events to be observed). Also implied is the analysis (actual or virtual) of the data stream into a more limited set of observations at times of interest. Let us call the data representation or image at a given time a *frame* and a set of successive frames a *series*.

5.2 DYNAMIC RESPONSE OF SENSORS

All sensor stimuli, without exception, can vary in time. Objects move, appear, and disappear. Illumination of objects is continuously changing, as are emissions from objects. All this adds up to a varying intensity observed by the sensor. Sensors must, therefore, be able to respond in a dynamic fashion to the stimuli to be of any use.

Let us return for a moment to one of the basic characteristics of sensors. Sensors convert the stimulus energy into a representation in an information system. For an idealized protosensor, this conversion takes place in an instantaneous fashion: an infinitesimal amount of energy can be detected. But actual sensors have energy thresholds and must perform some integration of the stimulus in time in order to detect it. Sensation, in other words, always requires a certain "window" in time to take place. This window may be very short, but it is still finite. This is even true when the sensor is measuring a flux, rather than a time-integrated and cumulative fluence. There is a trade-off. The longer the integration time, the more reliable the observation in a purely statistical sense. But a short integration time (e.g., a high sampling rate) will detect changes that would otherwise be lost to the information system. It can then react faster. (Long integration times are most suitable for objects that are dim and change very slowly. This is the normal state of affairs in astronomy, for example.) The integration time is the main limiting factor in the dynamic response of sensors. It defines the extent of the sensor in time. Different sensor materials will have very different response times to stimuli.

The dynamic response of the sensor also defines the frequency response[1]. We have seen that the bands of many sensor modes are frequencies: light frequency, sound frequency, and so on. But in such instances, the wave-nature of the stimulus is not directly perceived. The frequency, a parameter of the physical interaction that makes sensation possible, is translated into a quality, such as color or pitch, which is actually perceived. What is perceived are slower changes in intensity or the envelope of the wave (as in amplitude modulation). All sensors can respond in some fashion to a time-varying stimulus, which leads to a frequency response, even if of a macroscopic kind. Any of the parameters of the stimulus can change in time.

Complex sensors, as we have seen, have both extension and resolution (intension) in spatial field of view, in band, and in intensity range. This also applies to time. Sensor systems can have an extent in time as well as resolution time, leading to improved frequency response and sophistication in detecting temporal patterns -- that is, complex qualities in time. Human hearing is particularly sensitive to temporal complex qualities. And just as the mere fact of perception of spatial complex qualities implied the spatial extent of the object, so also do the temporal complex qualities imply a temporally-extended object. Moreover, temporal complex qualities requires memory on the part of the sensor or the information system.

Sensors themselves can adapt their sensitivity ranges to stimulus intensity, spatial field of view, and band in a dynamic fashion. This is considered below in section 5.5.

5.3 ANALYSIS OF THE DATA STREAM

Sensor theory assumes, to repeat, that the "real world" is a temporal continuity. It is not of itself broken up into discrete moments that the sensor (or information system it feeds) must synthesize or glue together. To the contrary, one of the main problems in the design of artificial sensors is the breaking up, the analysis, of the continuous data stream into a few times of interest.

In some rare cases, such as in the detection of elementary particles, it is not necessary to explicitly analyze the data stream, because the times of interest correspond exactly with the events detected. The stimulus, in other words, is itself one or more discrete pulses. The pulses define the data frames. But, as was pointed out above, this still requires continuous monitoring of the data stream by the sensor. Here, the data is sparse and the information system can handle all of it.

In most cases, however, the reverse is true. Generally, there is more information than the information system can handle or make effective use of. For a truly continuous data stream, there would be an infinite amount of information, not just cumulatively, but for any finite time span. The dynamic response or integration time of a sensor, as discussed in the previous section, defines the upper limit of the rate at which data can be taken. This is the *frame rate*. Suppose, for example, that we have an integration time of 0.001 second. This implies a maximum frame rate of 1000 per second. Thus, all actual sensors effectively break up the temporal data stream into finite data frames (even when the sensor response is analog, not digital). Figure 5-1 illustrates the analysis of the data stream into discrete frames.

There are thus two times to be considered here: the integration time t_i and the frame period t_f (= 1 / frame rate). Obviously, it is necessary that $t_i \leq t_f$. It is desirable that t_i be much less than t_f and that t_f be adjusted to the needs of the information system.

Television and other optical sensor systems typically have a frame rate of about 30 per second. Surveillance cameras may have a frame rate of only 1 per second. The human eye can respond (depending on lighting conditions) at about 10 frames a second: frame rates slower than this seem to flicker. On the other hand, sensor systems that are explicitly interested in waveforms, like hearing or radar, will have a very high effective frame rate. There is, once more, a trade-off between sensitivity to overall intensity (long integration time, low frame rate) and sensitivity to changes (short integration time, high frame rate).

A multimode sensor system must also consider the fact that the dynamic response and frame rate of each sensor type will be different. This is certainly true for human sense organs, for which hearing has a much larger and more sophisticated dynamic response than any of the other senses. (It is our most temporal sense. If vision provided us with this much temporal information, we would be overwhelmed by it.) There can even be differences within the same sensor. The rods in the retina react more quickly than the cones, and a computer screen which seems steady when seen directly (i.e., mainly by cones) can seem to flicker if seen in the corner of one's eye (i.e., mainly by rods). The fact that we can perceive the world in a single image from all the senses, each with a different dynamic response, is another proof that temporal continuity is in the world, and is not something put in by us. Indeed, it would be impossible to vary frame rates if the latter were the case.

5.4 TRADING TEMPORAL FOR NON-TEMPORAL

Many sensor systems deal with environs that change slowly in comparison to the response time of the sensors themselves. It may be advantageous in such a situation to trade time for better extent or resolution in field of view, band, or intensity range.

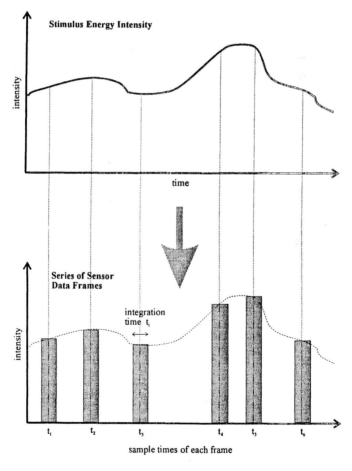

Figure 5-1. Analysis of the Data Stream into Frames.

One example is the representation of stimulus intensity in the sense organs of humans and animals by a frequency of discrete nerve impulses. High intensities are translated into a train of a large number of nerve impulses, while a low intensity leads to a correspondingly low impulse rate. In other words, intensities are represented by a kind of binary frequency modulation. The frequency of impulses is the carrier of information. The relation of impulse frequency to stimulus intensity follows the previously discussed Weber-Fechner or Stevens Laws. The important point that intensity at a single point in time is being translated into an extended pattern in time. This is possible because the maximum possible frequency response of the nerve fibers (several kilohertz) is far greater than the rates of change of biological interest (a few hertz). This is directly analogous to frequency modulation in radio, where the frequency of the modulating signal, such as voice or music, is much less than the frequency of the radio carrier wave.

An example of trading time for spatial extent is found in scanning. Scanning works on the assumption that the objects viewed are changing much more slowly than the scan rate. In scanning, as the name implies, a sensor covers a larger field of view by scanning its smaller field of view in time across a scene. The information is then integrated into a single image. Scanning is clearly akin to the process of integration of sensor elements in space that we saw in Chapter 3 for complex sensors. A temporal series of frames is fused into a single new image of wider extent. In scanning, there thus arises a distinction between the instantaneous field of view of the sensor and its total field of view: the total area scanned. Scanning is typical of both television and radar -- we are all familiar with the classic image of a sweep line going round and round on a radar scope. The image on a television screen is composed of scan lines. Scanning is also used in Forward-Looking Infrared (FLIR) sensors used by military aircraft to present an infrared scene of land and sky to the pilot. A single (expensive) infrared sensor element is used to scan a much larger field of view. Another example is found in space probes to other planets. Due to the long distances involved, and the corresponding feebleness of radio signals, images are transmitted back to earth very slowly, one pixel at a time, often requiring hours for a complete image to be constructed.

Scanning can take place on multiple levels. Human vision provides an example of this. We move our eyes continuously (saccades), often without being aware of it, to scan the world before us. Indeed, it is very difficult to hold one's eyes entirely steady (try it). Our eyes also move to respond to rapidly changing situations or to "track" objects. For more slowly changing situations that demand a change in the field of view, we can turn our heads. For an even greater change, we can move our entire bodies to change the visual field of view.

The question of time versus extent and resolution is an enduring one in sensor design. The faster a sensor must be, the more difficult and expensive it becomes to maintain good resolution, because larger sensor elements are more sensitive than small ones (i.e., they capture more energy). This can be seen even in photographic film. Fast films are grainy, while films of high resolution are slow and require good lighting conditions.

But here, once again, the requirements of the system or living being employing the sensor enter in. A sensor needs to be fast enough to deal with the pace of changes in its environs. If it is much slower than that, it will be useless. If it is much faster than that, it is either an unnecessary extravagance, or the door is opened to trading the available "unused" time for increased sensitivity range of intensities, increased spatial extent and resolution, or increased bandwidth.

5.5 Variable Sensors

In a *variable sensor*, the sensor attributes -- field of view, intensity range, band -- can be changed as needed. This was already touched on in the previous discussion of scanning. But the object of scanning is to increase the spatial field of view of the sensor. A variable sensor changes its parameters without joining a temporal series of images into one -- although a variable sensor can obviously be used for that purpose. The aim of a variable sensor is to adapt its field, range, and/or band to the best possible relative to the current situation. A larger effective extent is covered at less cost than a sensor that encompassed the entire extent at all times.

The human eye is a variable sensor of great sophistication. The iris expands and narrows to adjust to lighting conditions. The retina itself changes its sensitivity in low light conditions -- the familiar phenomenon of dark-adapted vision. The rods in the eye undergo a chemical change making them much more sensitive. (The cones do not participate in dark-adapted vision, which is why such vision is in black and white rather than in color.) Not only do the eyes move in their sockets to change and expand the effective visual field of view, their lenses flex in order to varying the spatial depth of focus.

Another example, unfortunately not belonging to human vision, is that of "zooming" -- the continuous change of magnification. This is a standard feature of telephoto lenses, for example, and can also be found in telescopes and binoculars.

The variation of the sensor band is simply that of "tuning" the sensor to a new set of stimulus frequencies, etc. Tuning is remarkably little used in biological sense organs. For artificial sensors, it most prominent in the sensing of electromagnetic waves. Radio receivers are literally tuned to different frequencies, while optical sensors can use a set of filters to select different light bands of interest.

Very often, a variable sensor will be designed to automatically modify its parameters in response to a changed situation. Such sensors are *controlled* or *automatic* variable sensors. Occasionally, these are called *intelligent* sensors[2]. There is a feedback between the intensity and other stimulus parameters sensed and the conditions of the sensor. Such sensors have their own information system -- some kind of prior information or knowledge, however primitive, is needed to account for the phenomenon of sensor self-control. Automatic sensors are thus not "pure" sensors, and their extensive description requires input from both control theory and robotics.

5.6 UNITY AND DIVERSITY OF THE SERIES OF FRAMES

The sensor output is thus a series of frames in time. These frames can be evenly spaced, reflecting an explicit division of the data stream into times of interest, or they can be irregular, reflecting detected events of interest. Or, the frame rate itself can be varied. (This is the case, for example, in the infrared sensor of a guided missile. The frame rate increases as it closes on its target.) The series of frames may be virtual rather than actual, depending on the temporal resolution of the sensor.

Each frame in the series will have an associated time. These times allow the information system to properly relate and order the frames. These times are not supplied by the sensor or information system; they are just the markers of the excision of the frame from the continuous input data stream. Thus, the objective unity of frames in the series comes from the continuity of the data stream itself, and not from the information system or sensor. Memory is required in order to hold the series, but the proper order of frames does not come from it. The unity of the temporal series within the information system is the memory of its unity in the real world. The diversity or differences among the frames in the series come from changes in time -- either of the sensor or of its field of view.

Our knowledge of changes will be limited by the effective frame rate. It must be emphasized, however, that changes in the frame rate on the part of the subject do not change the object known. As progressively higher frame rates are used -- more temporal resolution -- greater details of changes are discerned and there is a smooth convergence to the temporal properties of the world. The frame rate itself can be varied in a smooth and continuous fashion. It is assumed that as the frame rate goes to infinity, the changes between frames go to zero. (Truly discontinuous change does not occur physically. Finite movement must take place in finite time.) But it is not necessary to have an infinite frame rate to grasp changes of a finite complexity -- a fact expressed by the *sampling theorem*, which will be discussed further in Chapter 8.

CHAPTER SIX

ACTIVE SENSORS

6.1 ACTIVE SENSING IN GENERAL

An *active sensor* is capable of acting on its objects in order to obtain more information about them. It interacts with them, rather than just being affected by them, as in purely passive sensing. It not only receives energy from its environment; it beams out energy to the world and receives back that energy as reflected by objects.

It is important to realize that not just any combination of sensing and acting constitutes an active sensor, even when the sensing and acting belong to the same mode of physical interaction. There must be a coordination of the two in the detection of objects and their properties. The human voice in combination with hearing, for example, is not an active sensor; our voices do not interact with objects in a way that bring new information about them to our ears. Only in echoes from distant objects do human voice and hearing act jointly as a very crude kind of active sensing. (Of course, voice and hearing are very well coordinated for *social* interactions. But this is something that belongs to intelligence, not to sensation as such.)

Active sensing has two clear advantages over passive sensing alone:

1) It does not rely on chance ambient conditions in order to sense objects.

It can illuminate the object to "see" it under any conditions. An active sensor, in other words, makes the ordinarily invisible, visible.
2) Further, an active sensor can narrowly select what kind of illumination it will cast on the objects. Hence, it can isolate what features it can see. It can focus on exactly what it seeks to find, and thus overwhelm ambient background noise. It can sense just that energy it has sent out.

Clearly, an active sensor can make a much more sophisticated investigation of objects. It can focus on particular aspects, as well as discovering aspects inaccessible to purely passive observation. Unlike a passive sensor, an active sensor system can investigate all properties of the objects that fall within the sensor's mode, field, and band. Or, it can seek out just those few properties that the system employing the active sensor cares about. Active sensing is thus a way of *querying* the object: we ask the objects about their properties rather than waiting for the objects to tell us on their own. This means also that an active sensor is much more difficult to fool than a passive sensor. It can tear aside the camouflage netting to see what is behind it. The active sensor does not simply listen, it chooses what questions to ask.

Despite these notable advantages of active sensing, it has hardly been considered at all by epistemologists and other students of sense perception. The purpose of this chapter is to begin to remedy this lack. I hope to show that active sensing is a crucially important problem to all theories of sensation, precisely because the active sensor can interact with its objects. Sensation has always been treated as something strictly from the object to the observer, a purely one-way process of causation. Active sensing is two-way. There is a "closed loop" or circuit between the knower and the known. It is once more a process of querying the object, rather than simply observing it.

The classic example of active sensing is radar. A radar transmits a radio wave (either continuous or pulsed), almost always aimed in a narrow beam. The radar beam bounces off of objects, and the reflected radio waves are received back by the radar. The range is easily determined from the time the radio waves took to make the round trip from the radar to the objects. Examination of radar waveforms (e.g., Doppler shifts) can provide information about the movement of the objects. More sophisticated combinations of radar waveforms can give one important information regarding the identity of objects (e.g., what kind of aircraft is out there). In water, sound waves can be used like radar, to give sonar. Nature has come up with an active sensing system in echolocation by bats. It is remarkably similar to both radar and sonar. Bats not only use ultrasonic sound waves to locate objects and determine the range to them; they can also determine motion of objects by examination of the frequency spectrum of the reflected sound waves.

The only human active sense is touch, especially in the hands. We can reach out to touch objects, pick them up, and manipulate them. We can move our fingers over the surfaces of things. We can conform our hands around objects. Indeed, without the active feature of touch, it could only tell us when we bump into objects. But with the ability to move our hands and fingers to make contact with objects, touch becomes a way of discovering their spatial features.

Modern technology has vastly expanded the scope of active sensing, and with it the scope of possible human sensation. Radar is the best example of this, but even something as simple as carrying a flashlight in a darkened room is an example of active sensing.

Active sensing is also finding increased use in robotics. Robots can navigate by use of ultrasonic sonar (rather like bats) or by use of laser rangefinders. They can sense the shapes of objects by very clever means like shining patterned light upon them. There is nothing that prevents the use of instant holography to discern the shapes and distances of objects, although this technology is too difficult to be practical at present. The ultrasonic motion detectors used in some home security systems are active sensors.

Medicine is another important and growing application of active sensing. Ultrasound, CAT scans, and NMR imaging are all active, yet non-invasive, ways of sensing the state of the internal organs of a patient.

6.2 ACTUATORS

All active sensors, by definition, have *actuators*. To recall the definition given in the first chapter, an actuator is the reverse of a sensor. If a sensor translates energy into information, an actuator translates information into energy. The sensor is an interface from the outside world to an information system; the actuator performs the converse role of interface from the information system to the world. Sensors receive energy; actuators transmit energy.

An actuator generally seeks to cause changes in the state of objects in its environs. Just as all animals have sensors, they also all have one or more kinds of actuator: legs, arms, fins, wings, jaws, and so on. The human hand is not only an active sensor, it is a complex actuator of great sophistication.

The kinds and classifications of actuators run exactly parallel to sensors. Actuators are classified primarily by mode, and there are as many possible modes of acting as there are of sensing. All actuators can be represented as combination of infinitesimal *protoactuators*. Protoactuators, like protosensors, are characterized by the conditions of mode, method, spatial field, intensity range, and band. There are actual actuators that closely approximate a protoactuator, such as a pulsed laser. Finite width of field, range, and band yields a *simple actuator*. Most real actuators are complex actuators: they can be decomposed into simple actuators. When this decomposition is real rather than virtual, we have a complex actuator made of discrete parts -- a machine, in other words. Complex actuators can be constructed using the rules previously established for complex sensors. A combination of different kinds of actuators yields a *multimode actuator*. And just as there are variable sensors, so also there are variable actuators.

The form of the data representation input to an actuator (whether simple or complex) is identical to that for sensors. But whereas the sensor image tells the information system what it has received from the world, the data sent to the actuator tells it what to do. We may identify "qualities" in the information sent to an actuator by the information system, but such qualities will tell the actuator of a pattern it must transmit to the world or impress upon the objects. Clearly also the information input to the actuator defines the number of possible patterns or actions that can be carried out.

6.3 COMBINATION OF ACTUATORS AND SENSORS

An exhaustive examination of actuators and their properties belongs to robotics and not to the theory of sensors. Here our concern is how actuators can be combined with sensors to form active sensors. First of all, the mode, field, be combined with sensors to form active sensors. First of all, the mode, field, and band of the actuator will nearly always be designed to match that of the sensor. The purpose of the actuator in an active sensor is to add to the sensor's ability to see its objects. Thus, it must be matched to that sensor. The only case where that would not be true is when the object itself makes a significant transformation of the energy from the actuator. An example is fluorescence: one can use a combination of an ultraviolet lamp and a visual optical sensor to look for fluorescent materials.

To repeat, the actuator is coupled with the sensor to assist the sensor. The reverse combination is also possible: the use of a sensor directly coupled with an actuator to assist it. The best artificial example of this is a guided missile, where a radar or infrared sensor is used to keep the missile on target. In this case, one obtains a controlled actuator -- but this is once more a topic for robotics.

An actuator can be directly joined with the sensor or be separate from it. Let us call these *integrated* and *separated* active sensors. Figure 6-1 illustrates the difference between the two. In the former case, we have a transceiver. Which configuration is chosen depends a great deal on what is being sensed. An integrated active sensor will be able to measure only reflection from the object, and thus yields a reflection image. Radar is once again an example. A separated active sensor can measure the contrary aspects of transmission and absorption by the objects. This yields a transmission image. An example is the x-ray: we are interested in the x-rays that pass through or are absorbed by the object, not the tiny number of x-rays that are scattered or reflected. For this reason, an x-ray machine must have actuators separated from sensors. We can also imagine a combination of the two types, and have an active sensor with one actuator and two sensors. One of the sensors would measure reflection and the other transmission.

6.4 ACTIVE SENSORS AND KNOWLEDGE

The data received by an active sensor may not be inherently any different than that of a passive sensor. What distinguishes it is the ability to specifically query its object. The active sensor can isolate features of its environs for study and selectively analyze its objects. It is difficult or impossible for a purely passive sensor to do that. A passive sensor is forced to wait upon ambient conditions, such as natural lighting, or upon the objects themselves to transmit, as is the case with nearly everything we hear.

We now come to an important point. All scientific experimentation is a kind of active sensing. The whole point of experimentation is to hold certain aspects of the objects constant so that other properties can be examined. It attempts to isolate features of objects in a systematic way. Moreover, experimentation is an active process; it is not simply observation and data-taking. It attempts to discover properties that would be invisible to a passive observer. Experimentation, in other words, is a precise way of querying the object and analyzing its properties. It involves acting and sensing in concert. This is its connection with the theory of active sensors.

One of the few philosophers to understand this aspect of experimental science was Kant when he wrote in the preface to the *Critique of Pure Reason* that what we know certainly in natural science is what we ourselves put in[1]. Experimentation is querying, an active process to which the scientist contributes.

We saw in Chapter 2 that a protosensor could not be in error: whenever it registers "1", its stimulus is present. In the converse case of a protoactuator, it must produce a stimulus corresponding to its mode whenever it is signaled "1" by the information system. Now consider an *active protosensor* -- the pairing of a matched protosensor and protoactuator. For the purposes of sensation, the signal sent out by the actuator may be considered an *a priori* and for two reasons: its conditions (like that of the protosensor) are predetermined and its action always precedes the sensation. We always go from the protoactuator to the object and back to the protosensor. Moreover, we can always distinguish the effect our illumination of the object from the ambient background simply by switching the illumination off. The active protosensor does not observe only those objects "out there" emitting "its" specific stimulus. It can "reach out" to those objects independent of ambient conditions. It can initiate the relation of the sensor and the object, in other words. It can know with certainty if there is anything in its environs that interacts with its stimulus at any time.

It is interesting to note, too, that active sensation is the exact reverse of the

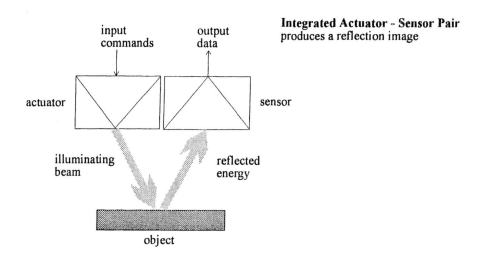
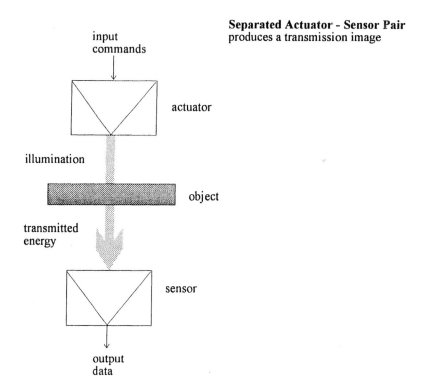

Figure 6-1. Integrated and Separated Active Sensors -- a Schematic.

normal process of cognition. Cognition typically goes from sensing to acting -- an extreme example of this is found in reflexes. But active sensing is acting in order to know rather than knowing in order to act. It is possible for sensing and acting -- both active sensation and acting for the purpose of changing the object -- to form a seamless whole. An example of this is when we manipulate objects with our hands. In using woodworking tools, or playing a musical instrument, or typing on a keyboard, there is such an integration of sensing and acting that it is impossible to tell where one stops and the other begins. They are fused into a whole. We actively sense with our fingers in order to know, yet this knowledge immediately issues into more action, which evokes more sensations, and so on. If we consider proprioception (the sense of the positions of the limbs), we see the same thing. There is a seamless integration of sensation and action, a closed loop.

Active sensing is not, however, a panacea for whatever ills an imperfect information-gathering system. It can have distinct drawbacks. Because an active sensor can be operated to sense only what it illuminates, the system employing it then sees only what it wants to see. It becomes blind to whatever it is not looking for. This disadvantage is the reverse side of the active sensor's capability to narrowly query objects. Active sensing also has the practical drawback that the sensing system may reveal its existence, its position, and its identity. That is because the active sensor must transmit energy, which can then be sensed by others. If one wishes to see, one also risks being seen. This is of great importance on the battlefield. When an air defense radar is switched on, for example, it reveals its location and becomes vulnerable to a radar-seeking missile fired by an opponent.

6.5 THE ACTIVE ARGUS SENSOR

Chapter 4 discussed the theoretical upper limit for sensors, a sensor combining all possible modes, with infinite extension and resolution of field of view, intensity range, and band: the Argus sensor. Now let us consider the addition of active sensing to the Argus sensor.

The active Argus sensor would have a mode of actuation corresponding in every way to its modes of sensation. It would thus be able to transmit any kind of physical energy, in any intensity, to any place. The active Argus sensor could thus interact in every possible way with physical objects. It could examine objects one aspect at a time and completely. The limits on the active Argus sensor would be only the limits of the information system that employs it, as the active Argus sensor could produce an infinite amount of information at each moment. In reality, the active Argus sensor is to sensor theory what the Turing machine is to artificial intelligence. It represents a theoretical ultimate.

The active Argus sensor would be capable of sensing everything that can be sensed, because it can physically interact with everything in the world. It would thus also yield everything there is to know about the physical world at a given time: it would provide *the materials for cosmoscience*. This is so, because everything in the physical world has at least some physical interaction, no matter how weak -- if it did not, it could not belong to the physical world. Indeed, the physical interactions are what define it physically: physical properties are possible relations. The active Argus sensor is capable of interacting in all physically possible ways. Thus, it can detect every property of everything. This means it provides comprehensive data about physical reality at any given time. Such data, if viewed by an intelligent being, will yield everything that can be known about the physical world; i.e., cosmoscience.

PART II

THE MEANING AND VALIDITY OF SENSORY INFORMATION

CHAPTER SEVEN

THE SENSORY OBJECT

7.1 GENERAL CONSIDERATIONS

Up to this point, we have considered sensation almost exclusively from the side of the sensor, while alluding only briefly to the objects the sensors observe. Only in the chapter on protosensors were the attributes of the objects discussed at any length. The exposition of the properties of sensors themselves is now complete. It is now time to change perspective to explore the object and the relation of its properties to the data representations produced by the sensor: in other words, the truth of sensation.

Such an investigation must begin with the *sensory object*, the "something" out in the world about which the sensor gathers information. The sensory object was touched on as early as Chapter 1 when the distal stimulus was mentioned. The distal stimulus is the bundle of physical interactions streaming out from an object. If the distal stimuli is integrated over all modes, the sensory object is obtained. It is the cause and source of the proximal stimulus -- the stimulus that actually irradiates or impinges on the sensor. The next chapter examines in depth the relation between the distal and proximal stimuli -- the problem of projection. But a necessary preliminary to that discussion is to understand as completely and clearly as possible the sensory object (distal stimulus).

The problem of the sensory object has two main points to address:

1) The representation (if possible) of the sensory appearance of an object by a combination of infinitesimal *proto-objects*, in analogy to protosensors and protoactuators.
2) The relation or mapping between the sensory appearance and the real properties of the object. Is such a mapping between the two consistent and reliable? Is such a mapping possible -- indeed, *are* there real properties behind the appearance?

Of the two points, the latter is the traditional problem of the reliability of sense-perception, because we do not grasp the inner natures of things merely from how they appear to human sense perception. The first point -- that of the proto-object -- is one of the keys to the latter, so we must consider it first.

7.2 Representing the Sensory Object through Proto-objects

In sensor theory, there are two ways we can consider any exterior object:

1) In a purely *physical* sense, as a combination of physical parts. Here, the object is analyzed into macroscopic parts, and these into molecules, atoms, and so on in a certain arrangement. This is the object's physical nature.
2) In a purely *sensory* sense, as a combination of sensory parts. The object will ultimately be represented in that way as a combination of atomic proto-objects. This is the object as a complete sensory appearance; that is, how it would appear to the Argus sensor.

The problem of all natural science is the reliable inference of (1) from (2). We thus need to examine what proto-objects are and how the sensory object can be composed from them.

All sensory objects have some location in time and space relative to the sensor. They will also have some extension in space and duration in time, even if it is an infinitesimal dx or dt. The sensory object is thus a spatio-temporal field, just like an image. It may always be represented as some state matrix $A(x,y,z,t)$. This also means that we can always decompose the sensory object into a spatio-temporal combination of atomic proto-objects (actually or virtually), according to some order.

The sensory object (hence also the proto-objects) are characterized by mode and band, just like protosensors. Since the mode defines the possible kinds of physical interaction, the object can only have one or more of these modes. Generally, an object will have some interaction for several different modes. The multimode integration of proto-objects will be on the basis of spatial collocation. For each mode of interaction, there will be the state parameters like frequency, polarization, and so on. This is identical to what we considered for sensors. Thus, the sensory proto-object is a spatio-temporal field of values, and at each point there is a set of parameters for each mode.

There will also be an intensity at each spatio-temporal point of the sensory object. But we must be careful what "intensity" means here -- it is not like the energy a sensor passively receives from without. Rather, for an object, there are emission, reflection, transmission, transformation, and absorption of energy. (Transmission means that the object is transparent to that kind of energy; transformation means that the object changes one kind of energy into another, as in fluorescence or black-body radiation.)

Objects can correspondingly be classified as *passive* and *active*. A passive object reflects, transmits, transforms, or absorbs energy. In other words, it is not "visible" unless illuminated from without, either by natural ambient sources of energy or by an active sensor. It does not emit energy on its own, just as the moon shines by reflection of sunlight.

There are also active objects, those that do emit energy on their own. Active objects may be subdivided into *natural* (or *uninformed*) and *intelligent* (or *informed*). A natural or uninformed object emits energy as part of a natural process, thus in a completely unintentional way, in conformity to natural law. Its isolated meaning cannot go beyond the natural level; it just reflects the order of nature. An intelligent or informed object, in contrast, emits a signal that intends to communicate something. The physical carrier of the interaction bears an encoded message, which need bear no relation to the physical properties of the object. The meaning of the interaction, in other words, goes beyond the physical level and cannot be reduced to it. No sensor, in and of itself, can decode what it means.

This means that a proto-object must have both passive and active aspects -- it must be like an active protosensor in reverse. A "protoemitter" is simply a protoactuator. A "protoreflector" or "prototransmitter" is a protosensor directly combined with an identical protoactuator -- as soon as it detects energy of its type, it immediately emits energy of the same type in the same or different directions. A "protoabsorber" is a protosensor alone (its information output need not be in use). Finally, a "prototransformer" is a protosensor combined with a protoactuator of a different type.

Thus, *every proto-object can be represented as a combination of a protosensor and a protoactuator*. Whatever energy goes into the sensory object requires protosensors. Whatever energy comes out requires protoactuators. And, to be seen by a sensor, every object must have some mode of actuation -- it could not emit or reflect energy otherwise. The intensity of the energy from the proto-object is thus:

$$O(t) = aI(t) + E(t)$$

where $I(t)$ is the intensity of energy from without as a function of time; a is the re-emission or reflection coefficient (e.g., albedo), and $E(t)$ is the object's self-generated energy emission. In other words, the energy intensity must combine descriptions of a protosensor and a protoactuator.

This brings us to an important point. For every proto-object, there is an active protosensor that matches it, and can know that proto-object completely on the sensory level. This must be true for the following reason. For every kind of protosensor, there is a protoactuator that perfectly matches it -- they are complements of each other. Every proto-object can be reduced to a combination of a protosensor and protoactuator, as we have seen. Suppose the protosensor is of arbitrary type A and the protoactuator is of type B. There is a corresponding active protosensor consisting of the reverse combination of protoactuator A and protosensor B. The active protosensor is, in other words, the mirror image of the proto-object, the "key" to the proto-object's "lock". The active protosensor can apprehend its corresponding proto-object completely. By simply observing, it can detect the E(t); by interacting with the object, it can determine the coefficient a. It should be noted, however, that the sensor cannot understand by itself the pattern of E(t) from an intelligent emitter.

Since every sensor is representable by field of protosensors and every sensory object is representable by a field of proto-objects, it is possible (in theory) for sensory objects to be known with complete reliability. The sensory object is a source or reflector of physical energy of the various modal types -- it is an energy flux, in other words. The sensor is the corresponding energy sink. A perfectly consistent and reliable mapping is possible between the object and the sensor. For any sensory object, we can devise an active sensor capable of extracting every one of its properties. This is why in the previous chapter it was stated that the active Argus sensor would provide the material for cosmoscience -- it would literally extract all the physical properties of objects there are.

7.3 CORRESPONDENCE OF SENSORY AND PHYSICAL OBJECTS

We have attempted to show that a sensor can, in theory at least, completely and reliably grasp all the attributes of the *sensory* object. There is a "mapping" between the two. Now we come to the more important question. It is possible to proceed from the sensory object to the real, physical object and its nature? Is there a knowable and necessary correspondence between the physical properties of an object, its nature, and its modes of physical interaction? Is there, in other words, an invariant mapping from the sensory object to the nature of the physical object? Locke saw this problem clearly in his useful (yet little used) distinction between "nominal essences" and "real essences". The nominal essence is the sensory appearance: it is how we describe a spatial pattern of colors, brightness, sounds, etc. This corresponds to our sensory object. The real essence is the actual, physical structure of the object, which Locke (under the influence of the chemist Boyle) supposed to be atoms in various spatial arrangements. The whole aim of modern science is, of course, to discover what the physical "real essences" of things are. The question for sensor theory is, *can* we pass from the "nominal essence" to the "real essence" and, if so, *how*?

This is a question that has vexed philosophers for centuries, and I do not purport to solve it here. My aim is to point out the ways in which general sensor theory can clarify, simplify, and provide an orderly approach to it. It is important to note here that, when speaking of the mapping between the properties of the object and the sensory appearance, that this is at the complete (i.e., infinitesimal) level. In other words, it is as if the object were viewed by an active Argus sensor.

First, all physical objects must have some mode of physical interaction, and hence of emission, absorption, or reflection. This means that for every physical object, there is a corresponding sensory object (a spatio-temporal bundle of interactions). There can be no essentially invisible physical things, no "stealth objects". Conversely, if something cannot interact physically in some way, then it is not part of our physical universe, and can have no sensory object. Thus, whatever appears to a sensor has some physical manifestation, and vice-versa. Whenever something is sensed, there is a physical object "out there", for, by definition, a sensor responds only to physical stimuli. We dealt with this problem previously for the protosensor.

Second, the sensory object contains the *complete* set of the physical object's interactions. If we were to consider only a subset of the object (as human senses must, for example), ambiguities are immediately introduced. For example, suppose we see a circular shadow cast upon a wall. One might immediately jump to the conclusion that there is a circular object "out there". And one would be partly correct -- there must be something round that casts the shadow. Yet, at the same time, there is an irreducible ambiguity, as there are in fact an *infinite* number of three-dimensional objects that could cast a circular shadow: an opaque circle, a sphere, any kind of spheroid or ellipsoid, any cone or cylinder viewed on end, and so on. Yet the sensory object, as viewed by an Argus sensor, would not have this problem, as it would view the entire three-dimensional field, rather than a two-dimensional projection. Consider another example. I hear loud music coming from a nearby park. Is it a live band or is it a recording? The auditory object could be the same for each, thus it could be either one. But the auditory object is only a subset of the sensory object: from the complete sensory object, one could tell the difference. The next chapter considers this question at more length.

Third, since the sensory object possesses all possible modes of physical interaction with an object, a unique mapping of the physical properties of the object into the sensory object must exist. The sensory object need not appear the same as the physical object, any more than the internal structure of an atom, for example, resembles its spectrum. What is necessary is that there is a *unique and invariant mapping* between the two. It is true that many different objects will interact with the same physical carrier (e.g., light) and that the same physical object will interact with several different carriers (e.g., an object is both visible and heavy). Nonetheless, the spectrum and pattern of interactions will be unique for each kind of object. Every physical substance has a unique "signature" of physical interactions as it were, corresponding to its properties. The sensory object thus permits exact identification of physical objects. Indeed, the properties are simply what connect objects; they are the termini of interactions. To return to the example of atoms, although atomic spectra do not resemble the structure of atoms, they were the means by which that structure was discovered. The energy spectrum of each kind of atom or molecule is its "fingerprint"; it is not possible that any other kind could be mistaken for it.

Every property is a potential relation. By looking at every possible physical interaction, we thus discover all physical properties. If there are no possible interactions, there are no properties; conversely, the richer the interactions, the more complex the properties are. Every different kind of physical object will thus have a unique set of properties and a unique spectrum of physical interactions -- a unique sensory object. Thus, there is a mapping between the physical object and the sensory object. And thus it is possible to discover the physical natures of things from their sensory objects -- if one could only view the sensory object in its entirety! With increasing sensor resolution, the sensory object will converge to a perfect expression of the real physical properties. It is an underlying assumption of science that such a convergence is theoretically possible.

The above discussion does not entirely solve the problem of the relation between the real and sensory objects; it only shows how sensor theory can clarify it. We have thus far shown there is a reliable mapping between the sensory and physical objects: the distal stimulus expresses the physical properties of the object. This will satisfy scientists and engineers, but not the philosophers. There remains the question of whether we can attain reliable knowledge of objects from the partial spatial images that we are always actually limited to. This will be considered in the next chapter. There is also the question of whether the natures of things are invariant. This is, of course, a working assumption of natural science; if there were no constant natures of things to be discovered, science would hardly be worth pursuing. This question goes far beyond sensor theory and is hence not answerable here. It should be noted, however, that -- from a purely physical point of view -- something's nature could not change in isolation. Physical properties, being potential relations of various sorts, form a system. The whole system must change if one part changes; thus, the fundamental properties of one object could only change if the laws of the entire universe were also to change. This means that the properties measured by sensors will only change if they change everywhere. And if an object suddenly sprouted entirely new properties, these would have to find some way to relate to other objects, or else they would remain physically irrelevant.

7.4 INTELLIGENT EMITTERS

The intelligent emitters discussed in section 7.2 raise the possibility of the sensory manifestation of something that is, of its inner nature, suprasensible or immaterial. This is obviously a philosophical, and not a scientific, question. What we are dealing with here is something akin to the Kantian thing-in-itself or noumena, which causes sensations in us, but is otherwise completely unknowable by our empirically-directed minds. One could not infer the inner nature of the Kantian thing-in-itself from its sensory appearance, any more one could deduce the structure of a television station simply from the image that appears on a television screen. Now this is just like what we have described for the intelligent emitter -- there need be no connection between the emitted pattern and the *physical* nature of the emitter.

The relevance of this to sensor theory is twofold. First, it is assumed that the nature of a purely *physical* object can be known from the sensory object, as it contains all the relations to other physical objects, thus everything that is physically relevant. If there is a non-sensory object behind this, it is in fact physically irrelevant, cannot have sensory appearance, and science can't possibly care about it.

Second, a non-sensory object must have a nature or essence that is more complex than, is above and beyond, the sensory object in which it manifests itself. Indeed, sensor theory implies that if there are non-sensory objects, they are intelligent. The converse is also true: the intelligent can never be entirely reduced to the sensory and the material. This is of relevance to the mind-matter (or mind-brain) problem. The sensory object defines, as it were, the lower limit of the real nature -- it is at most the real nature "seen in a glass darkly". Exactly how a non-sensory or immaterial object could have a sensory manifestation is not knowable through a sensor (and hence not amenable to treatment by natural science).

Actually, the distinction of sensory and non-sensory is alluded to in the definition of a sensor itself. The sensor acts as the interface between the physical world and an information system. Information, an encoded pattern, is not of its essence material. Although it may always require a material carrier in our world, it in no way depends on this carrier for its meaning. A word remains the same word whether written on paper, engraved in stone, or presented on a computer screen. An intelligent pattern of emissions, as we have seen, need bear no likeness whatsoever to the physical nature of the emitter. A purely physical object, however, can be represented as a combination of sensors and actuators whose relations are completely fixed. Thus, there can be a sensory manifestation of something non-sensible only if there is an information system feeding into actuators. The nature of the information system can only be known from decoding the signal it is transmitting through the actuators; it is not knowable in the way a physical object is, simply from the bundle of physical interactions that present themselves to the sensor. In other words, sensors, sensory information, and sensor theory, can never go beyond the physical *in and of themselves*. Something more is required -- an information system can only be understood by another information system.

CHAPTER EIGHT

PERSPECTIVE AND ERRORS IN SENSATION

8.1 FROM DISTAL TO PROXIMAL STIMULUS

Even when the connection between the sensory object and the properties of the physical object has been established, there remains the question of the relation between the sensory object and the stimulus received at the sensor -- the relation of distal and proximal stimuli, in other words. Exactly how the stimulus is transported from the object to the sensor is a problem central to the validity of sensation, because the proximal stimulus may *appear to us* to be quite different from the distal stimulus. If I see a house across the street, for example, I am seeing only a part of the house -- I do not see it from all perspectives simultaneously. Nor does the proximal stimulus necessarily possess all the richness of sensory modes that the distal stimulus does. The image is less than the reality. And it is a fact we occasionally make mistaken inferences from our perceptions. How, then, can perception (which necessarily perceives the proximal stimulus) be valid or veridical?

The relation of the distal and proximal stimuli is, in fact, the cause of most illusions and errors. These illusions are the source of the stock arguments of skeptics and idealists against the possibility of veridical sense-perception. A realist or empiricist must show, conversely, how this relation does not negate the truth of sensation, even if it might render it vaguer and less precise. One point that must be kept in mind throughout this entire chapter: sensor theory examines the relation of object and sensor for *all* possible sensors. To point out an illusion for *human* sensation (a particular case) does not mean it is a universal illusion, necessarily affecting all sensors.

There are three aspects of the transport of the distal to the proximal stimulus that must be considered. The first, that of *perspective* or *projection*, has already been alluded to. Perspective accounts for the fact that our view of an object is usually partial, and may be distorted, as in a fun house mirror. Section 8.2 investigates how perspective changes our knowledge of the object, and whether it can negate that knowledge altogether.

Perspective and projection still treat the relation of the object and sensor as if the stimulus were traveling through an empty space or ideal medium. In reality, sensors often operate in environments that are far from ideal. Driving in the fog is more difficult than driving in clear daylight. The medium through which a stimulus passes affects it, corrupting any sensor image and introducing noise.

There are also errors introduced by the sensors themselves. Up to this point, we have treated sensors as if they were constructed of perfect materials and manufactured completely without defects. Actual sensors, of course, are made of imperfect materials and vary from the ideal blueprint. Sensing in the "real world" necessarily involves errors coming from the sensor itself -- the image will be degraded from what is theoretically possible.

The nature and sources of errors in measurement and in communications have been extensively studied; as such, they are not a branch of sensor theory. An *error* for sensor theory is that which reduces the fidelity of the image to the object, beyond the inherent sensor conditions. An error either (1) adds something to the image that is not from the object or (2) removes something that should be in the image. Errors may be either *systematic* or *random*. A systematic error follows a pattern or law; it is essentially an unwanted signal in the system, like the presence of 60-cycle hum in a phonograph output. The pattern can be understood and predicted, however. A random error, on the other hand, is completely unpredictable in detail. Hiss and static are examples of random error. *Noise* is a common term for additive random error. Actual sensory processes will invariably have both systematic and random errors.

The real question for sensor theory is, once more, do the errors that necessarily arise in sensation in the "real world" cancel our knowledge of the object or merely degrade it in a well-understood way? Is it possible to correct for errors and recover what we really want to know from them? This is, in essence, the "signal to noise" problem -- how do we extract the signal from the noise? Indeed, how do we distinguish the two at all? Sections 8.3 through 8.5 deal with these problems. The chapter will last of all briefly discuss the relation between the sensor and the object as a communications channel with a certain information bandwidth. (Please note that errors and illusions *within* the information system itself will be dealt with in the next chapter. This is in keeping with the distinction between sensation and perception.)

8.2 PROJECTION, PERSPECTIVE AND STIMULUS DYNAMICS

The problem of perspective arises because of sensor limitations, which permit the viewing of only a portion of the sensory object. The sensory object (a bundle of physical interactions) will be projected out into the three spatial dimensions, will cover several modes and a large bandwidth, and may have a large range of intensities. Only the Argus sensor could capture the full set of interactions with all objects -- it has the comprehensive point of view. Only its images can contain all possible sensory objects. Other sensors must, at any given time, select a portion -- usually a small portion -- of the sensory object to view. A few strands of the entire bundle of interactions with the object are separated out by the sensor. Projection is not simply with regard to spatial extent and dimension; it also affects band and resolution.

This means the object is viewed from a restricted perspective, and such a view gives only *relative* information about the object. There is always a potential loss of information in this. The complete view of an object can only be reconstructed from a temporal series of frames from different perspectives -- but that assumes that the object is either unchanging or changing relatively slowly. What we perceive, then, is almost always something less than the object we claim to perceive. This fact alone -- the "partitivity" of sense-perception -- has been used to deny that we *directly* perceive objects[1].

The human senses, as one might expect, all give perspectival information in normal circumstances. Each of our eyes provides a two-dimensional view of a three-dimensional world; the depth perception of both eyes together is vague compared to the resolution in the other two dimensions. Vision is sensitive only to a single octave of electromagnetic radiation, and it reduces frequency information to mixtures of three primary colors. Hearing is more sensitive to the frequency structure of sound waves, but its spatial information is based on two points, the location of the ears on the head. Touch is the only one of our senses that can avoid perspective in space, because the hands (as well as the tongue) can conform themselves around a three-dimensional object. But if the object is large, only a small portion of can be felt at any given time. The reduction of information that occurs in smell and taste is even more severe. It should not surprise us that human senses can at times lead to illusions and faulty inferences.

Thus, except for the Argus sensor, the sensor image will be some subset of the sensory object: image $\subseteq$ object. The image of an object can never exceed the information contained in the sensory object. Or, put in a slightly different way, the sensor conditions intersect the sensory object at certain points, rather like the intersection of a line with a plane: image = sensor conditions $\cap$ object. The whole problem of perspective is to determine the projection of the sensory object onto the sensor. This itself has two steps: the transport of the distal to the proximal stimulus, and the reduction of the proximal stimulus to the sensor conditions. This requires a generalized optics or study of stimulus dynamics.

8.2.1 GENERAL KINDS OF PROJECTION

Actual projection of the distal into the proximal stimulus is mode-dependent, because the transport of each mode of stimuli is different. Light and other electromagnetic waves travel at 3×10^{10} cm/sec in a vacuum. The speed of sound (material compression waves) will depend greatly on the medium in which the sound waves travel. Chemical stimuli, which act on taste and smell, spread by a process of diffusion and advection in air or water. In some cases, we can actually observe stimuli being transported, as when wave rings on the surface of a pond spread from the point where a stone was tossed, or in the dispersion of a drop of a chemical substance in a glass of water.

The differences in the propagation of each modal stimulus mean that the study of stimulus dynamics divides into several specialized fields. Visible, infrared, and ultraviolet light are treated by optics. Sound waves are considered by acoustics. A special area of study could be established for every possible sensory mode. Despite the differences, however, it is possible to identify aspects of stimulus dynamics and projection common to all modes.

All projection is a transformation or reduction of the sensory object. These transformations are primarily due to conformity to the sensor's conditions (field of view, band, and intensity range), but also to possible geometric transformations, such as reflection, which may occur either in the medium or in the sensor. Errors due to the medium as such are not considered here, but in section 8.3.

There are four basic kinds of projection, common to all possible modes of sensation. These are: cropping, distortion, rebinning, and time lag. Any or all of the four can occur together. They are the four basic "dimensions" of a generalized understanding of perspective.

Given a description of these four for a sensor at a given time, we can map the sensory object into the sensor image.

Cropping is limitation or reduction in extent. The sensor typically views only a limited portion of the object, just as a map of the United States is a small portion of the entire globe. A complete picture can be reconstructed only from several different images. The reduction is defined by the sensor's spatial field of view, its mode, its band, and its intensity range. Cropping can also occur due to the kind of illumination cast on an object, as when a green object appears dark gray when a red light is shone on it. Reduction of the extent of the image also results from the fact that we normally only observe the surfaces of things and do not discern their interiors (this is related also to obscuration and attenuation -- see section 8.3).

Cropping occurs because actual sensors do not have unlimited width of field or bandwidth. Spatially, it means viewing the object from a particular perspective. In mode, it means limitation to one kind of interaction. In band, it means only a limited portion of that modal interaction -- a segment of the entire spectrum -- can be viewed. There is still, however, a potential one-to-one mapping between the features of the partial sensory object as viewed and the sensor image. Limitation to one small area of space or band is, in fact, usually desirable -- one typically wishes to attend to certain objects and exclude others.

The second kind of projection is *distortion*. This typically arises from the translation of a multidimensional object into an image of lower dimensions. The best example of this is found, once again, in map projections. If we make a map of the earth, its features will necessarily be distorted by the map projection: this is inherent in the reduction of the spherical surface of the earth to a flat piece of paper. And the distortion will not be the same everywhere in the map. In a Mercator projection, for example, distortion is zero at the equator and infinite at the poles. If distortion is minimal, isomorphic images can still be formed. Distortion will also occur from reflection, which reverses the image, and refraction, which may invert it. Reflection and refraction in the medium are considered below in section 8.3; but sensors themselves may contain mirrors and lenses. Distortion occurs in human vision, for example, when a three-dimensional object is projected into a two-dimensional and inverted retinal image. This means that visually-perceived extents are actually angular extents, leading to the illusion of a nearby person appearing larger than a distant building. Distortion and cropping have been exhaustively studied in geometry and optics, and their technical aspects need not concern us further here.

The third kind of projection is *rebinning*. Just as actual sensors do not have infinite extent, neither do they have infinite resolution. There are always a limited number of increments in intensity, band, or spatial field -- a finite sample. This leads to rebinning, the mapping from higher to lower resolution. (This is the same as the reduction of extent in the transform space.) Rebinning means that a one-to-one mapping of the sensory object into the sensory image may no longer be possible. The effect of rebinning is to "blur" or "smear out" an image. Indeed, we can never discern detail of the object finer than the sensor resolution -- this is a problem dealt with by sampling theory. Rebinning is quite evident in band for human senses, which splits the entire light spectrum into three primary colors, the infinite varieties of chemical substances into a few taste and smell qualities, and so on.

The final variety of projection is *time lag*. Time lag is due to the fact that physical stimuli propagate at a finite velocity. It takes time for the stimulus to reach the sensor from the object. This means that when we perceive an object, we see it not how it is, but how it was. Moreover, different physical carriers travel at different speeds. One common example of this is when we see the flash of a lightning bolt, but only hear the thunderclap a few seconds later. Light travels far faster than sound. Time lag is also due, once more, to the sensor's conditions -- the sensor is at a finite distance from the objects viewed.

8.2.2 PROJECTION AND THE TRUTH OF SENSATION

Let us return to the central question here: does the projection of the sensory object into the (limited) sensor image give us real and reliable information about the object's properties, whether directly or indirectly? Or, to the contrary, do limitation and perspective make knowledge of the object impossible? Sensor theory shows that projection *reduces*, but does not *negate*, the knowledge we can have of objects. This follows from the nature of the four kinds of projection: they are all transformations or reductions of the sensory object. Thus, the sensor image always contains some portion of the sensory object, however limited or distorted. Partial knowledge of the object is thus possible.

A swarm of objections can be raised against this immediately. We have already recognized that projection is the source of many, if not most, sensory illusions. By "illusion" here is meant something appearing to be what it is not. Mistaken inferences are always possible on the basis of our limited human sense faculties. Only the Argus sensor is immune from projective illusions, because it has the comprehensive view. For any finite sensor, projective effects mean a perspectival view and the possibility of illusion. Argument from illusion and mistaken inference would thus seem to negate the possibility of valid sensory knowledge. And skeptical and idealist philosophers through the centuries have made such arguments from projective illusion the centerpiece of their criticisms of sensory knowledge. Projection means sensory knowledge is relative, and because it is relative, it must be subjective. Then it is argued that subjectivity and illusion in turn mean that we never directly perceive material objects; what we perceive is a subjective intermediary: a sensation or sense-datum.

Yet projection does not mean that sensation is subjective, and there is no necessity to introduce intermediaries or deputies -- at least not for this reason. The theory of projection in fact solves *all* exterior illusions. This can be seen from the common varieties of exterior illusion, which are all examples of projection[2]. For example, when a coin seen at an angle appears elliptical or a distant tower appears to be the same size as a nearby man, these are examples of distortion resulting from the geometric projection of three-dimensional objects into two-dimensional retinal images. Reflections reverse the orientation of the image, but, again, this is simply a transformation of I(x,y) into I(-x,y). There are also illusions due to rebinning, as when there are apparent color changes under different conditions or when a shiny disk (even of a non-metallic substance like glass or mother of pearl) might be mistaken for a silver coin.

There is also the famous (or infamous) illusion of the bent stick. A straight stick, when partly immersed in water, will appear to have a sharp bend at the water's surface. This is an effect of refraction and thus due to the medium through which the stimulus passes, but its explanation is nonetheless one of geometric projection. How do we distinguish the illusory perception of a bent stick from the veridical perception of the "real" straight stick? It is straightforward, and it shows that the illusion is simply a geometric transformation that presupposes the straight stick to start with. First of all, the immersed stick does not appear bent to all sensory modes, which I can easily confirm by feeling the stick. Second, the illusion never appears in the absence of the refractive medium (it does not occur in a vacuum). Moreover, the amount of bending is relative to the kind of medium introduced: a stick half-submerged in alcohol or mineral oil will appear to be bent at a different angle than if water is used (i.e., different media have different refractive indices). That these effects are not subjective is proven by the fact I can take a photograph of them. Finally, the amount of bending varies with the sensor's angle of view in a lawful way-- if the stick is viewed end-on, for example, the bending disappears and the stick appears straight once more. In other words, *the illusion is relative to the sensor's conditions*. In these ways, the illusion is known to be an illusion and the projection of the real straight stick. It is a transformation, which can be corrected for exactly if the refractive index of the medium and the sensor's viewing angle are known.

Common exterior illusions are all relative to the limitations of the human senses. Exterior illusions are mode and sensor dependent: *there are no general exterior illusions*. For the Argus sensor, as previously mentioned, there would be no projective illusions at all, as it can view the entire sensory object. And if one holds that there is no *essential* difference between illusory and veridical perception -- both are purely subjective -- then how could the very distinction of the two ever arise? It would be completely arbitrary. The theory of projection explains exterior illusions completely and does not lead to such absurdities. Exterior illusions are due to the transformation and reduction of the sensory object to the sensor's conditions. These transformations, which obey laws, can be corrected for if they are known, either quantitatively or on the basis of experience. From the bent stick example, we can see the basic ways of correcting for projective illusions: the use of multiple sensory modes, multiple perspectives on the object, varying the sensor's conditions, and, best of all, active sensing. Indeed, those who argue from illusion against sensory knowledge have never, as far as I am aware, considered active sensing.

Further, we can continuously vary our parameters of view of the object, but the object itself does not change, only our perspective. We can move nearer or farther from the object, use a greater or lesser resolution, expand or contract the number of sensory modes being used. The appearance of the object begins to converge to a limit the closer the sensor approximates an Argus sensor. At the limit of the Argus sensor, the entire sensory object will always be grasped. Certainly, there is necessarily a potential reduction of information involved in any projection, unless the sensory object is completely proportioned to the sensor like a glove to a hand. (This can occur if the sensor is designed to view a specific kind of object.)

Projection thus means that information regarding sensory objects is partial, but it is not nothing. We still always grasp some portion of the sensory object. Projection of the sensory object into the sensor image still puts us in relation to the properties of the physical object. Sensory knowledge admits of degrees; it is not an all or nothing proposition. This accords well with common sense. Our everyday experience shows us that human sense-perception is at least partly veridical, for we are able to attain in action some fraction of what we intend to. The possibility of exterior illusions does not disprove that fact.

At the same time, the reduction of information inherent in projection introduces ambiguities. This is in keeping with the assertion that sensory knowledge admits of degrees: projection makes it vaguer. To return to an earlier example, a round shadow on a wall could be cast by any number of possible three-dimensional shapes: disks, spheres, cylinders, ellipsoids, cones, etc. The finite sensor's grasp of an object thus provides information for placing it in a class, rather than identifying it completely. At the same time, although the resulting sensor image is incomplete, it is still reliable in what it does tell us. The object that casts a round shadow must have a circular aspect -- it could *not* be a cube or a pyramid. A circle is necessarily contained in its topological make-up. If I see a round shadow, and then say, "I see a round object", I have spoken the truth: my perception delimits a class of possible objects, one of which is actually present. Thus, the reduction inherent in all projections does not destroy the possibility of knowing the physical nature of object; it merely makes it more general and vague. Instead of saying that the object is located exactly at point x, we now say it is located between the limits of x_1 and x_2.

And it is worth keeping in mind that physical objects are, after all, finite in both extent and complexity. We *don't need* an infinite field of view to see a human face or record a bird call. Infinite resolution is likewise unnecessary, as a finite object contains finite information. This fact is expressed by the *sampling theorem* used in signal processing[3]. To reproduce a continuous signal of frequency bandwidth W requires a sampling rate of 2W or greater. It doesn't require an infinite sampling rate. In other words, rebinning -- reduction in resolution -- need not remove any information about the object.

We do not need an Argus sensor to obtain knowledge of the world around us. Our unaided senses can tell us something, at least enough for us to live our lives. The human senses exist for the natural, pragmatic needs of human life: the partial knowledge they yield is adequate to identify different kinds of things. But the human senses are not adequate to obtain the complete information about objects that science requires: hence, modern science has advanced by use of scientific instruments and artificial sensors that overcome the limitations of the human senses.

8.3　Errors from the Medium

Physical interactions will nearly always have to pass through some medium on their way from the object to the sensor. This medium may be continuous, like air or water. Or it may be full of many obscuring, reflecting, and emitting objects, like the surface of our world is. The medium unavoidably influences the interaction of sensor and object, and thus introduces errors both by adding information we don't want and blocking the information we do want. The medium may, in fact, completely block knowledge of an object. The difficulties of dealing with errors from the medium stem from the fact that (unlike the sensor), we do not have knowledge of the *state* of the medium before sensing. The medium is continually changing, and it (like the object itself) must be sensed to be known.

Errors due to the medium can be divided into several classes:

(1) *Obscuration* (or occlusion) means that some object, opaque for the mode of sensor being used, blocks contact with the desired object. This is an everyday experience: we can't see through walls. Obscuration may be partial or complete. Partial obscuration -- that is, masking -- is capable of significantly changing inferences regarding the shape of objects.

(2) Closely related to obscuration is *attenuation*. Indeed, obscuration is nothing more than complete attenuation. Attenuation is the decrease in the intensity of stimulus, usually as a function of distance traveled in the medium. One of the best examples is the attenuation of light in the fog: nearby objects may be relatively clear, but at greater distances, things fade away into the mist. There will nearly always be some attenuation of the interaction between the sensor and the object. This is even true of outer space, where interstellar dust clouds make the stars behind them appear dimmer than they actually are. Attenuation can particularly lead to errors regarding the estimation of distances, because it changes the brightness of objects, a primary cue for spatial depth.

Both obscuration and attenuation are subtractive. They either remove part of the sensory object completely (something similar to projection), or fade it out. If the medium is structured -- turbulent rather than smooth -- the effects of attenuation or obscuration is to introduce random noise.

Obscuration and attenuation will act in different ways on different modes and bands of stimuli. (Just as there are no general illusions, there are also no general obscuration and attenuation.) For example, the medium may block certain frequencies of light and thus change the apparent colors of the object viewed: mountains, for example, that appear bluish or purplish at a distance may actually be clad with green forests. The same sort of selective filtering occurs in the atmosphere with respect to sunlight: much of the ultraviolet and infrared light is absorbed, while the visible frequencies are allowed to pass through nearly unaffected. The setting sun appears red because the shorter wavelengths of light are blocked by the greater thickness of atmosphere near the horizon. Low frequency sound waves will travel further than high frequency ones, modifying the audial appearance of objects at distance.

(3) *Shadowing* has effects similar to both obscuration and attenuation. Shadowing means the obscuration of the source of illumination -- whether natural or artificial -- of the object. It thus has relevance mainly for objects that are reflectors rather than emitters or in cases where active sensing is being used. If shadows are structured and/or moving rapidly (think of the shadows cast by clouds on a windy and partly sunny day), they become another source of random noise.

(4) *Deflection* is another main source of errors from the medium. Deflection means the dislocation of the interaction between the sensor and the object from what it would be under ideal conditions. For example, a light ray will travel in a straight line in ideal conditions, but in a medium, it will be scattered, refracted, and so on. Spatial deflection is the most familiar to us, and includes such phenomena as blurring, refraction, reflection, and diffraction. All these change the path along which the interaction propagates, and hence the apparent location and shape of the object. Deflection is the source of many optical illusions, and the more spatially acute the sensor or sense organ is, the more this becomes a problem. Deflection occurs not only in space, however. It also has relevance for time, in that a medium can slow down the propagation of the interaction. It can also be found in band; the medium can be fluorescent, for example, transforming the frequency structure of the stimulus.

Deflection can also introduce sensory objects or appearances that are not really there. It is the source of mirages. Multiple reflections can split the relation of sensor and object up along several paths, and there appear to be several objects rather than one -- "ghosts" on a television screen are an example of this. Because deflection is not only subtractive, but can add unwanted information to the appearance of the object, it is one of the most difficult of all error sources to deal with. Multimode or active sensing may become necessary to clear up ambiguities.

Deflection, like the other errors, can have a random character. Random deflection is responsible for the twinkling of stars. Such random deflections can seem to create objects than spontaneously appear and then disappear. If we are familiar with these fluctuations (sun sparkling on a lake, for example), they can be easily relegated to the background. Otherwise, such scintillations can become a major source of false or unidentified objects.

(5) A final source of errors is *emission* from the medium itself and the objects it contains. Again, this a familiar problem: trying to listen to a conversation over the roar of a crowd or of traffic is an example of this. Emission is additive, and its general effect is to drown out the desired signal, just as we cannot see stars in the daytime due to the brightness of the sun and skylight. A great deal of random background noise is due to emissions from the medium.

8.4 ERRORS FROM SENSORS

Errors arise, to repeat, from two main sources: (1) imperfections in the sensor itself, including secondary modes of sensation, and (2) the influence of the medium between the sensor and the object. As we have just seen, the inherent limitations of sensors in spatial field of view, band, and intensity range are not in themselves a source of errors. They reduce the amount of information available about objects, but they do not corrupt it. Projection and perspective, correctly interpreted, do not yield erroneous information about the object.

8.4.1 MATERIAL LIMITATIONS

Sensor defects are due primarily to the imperfections of the materials used in their construction. The engineer (in the case of artificial sensors) and Nature (in the case of biological sense organs) are confronted with a limited number of possibilities for sensor components. Physical materials, to say the least, do not give the sensor designer a nice selection of protosensors from which to build complex sensors. The sensor material must, on the one hand, be capable of transduction. Transduction for nearly all sensors means the transformation of the stimulus into an electrical (or electrochemical) signal. This means the sensor material must be sensitive to the influence of its desired stimulus. On the other hand, a useful sensor material must be limited in its sensitivity to specific kinds of stimuli. It must respond over a wide range of intensities in a consistent way. Further, the sensor material must not be exorbitantly expensive, or extremely bulky, or require special treatment for operation (e.g., cryogenics) if it is to be

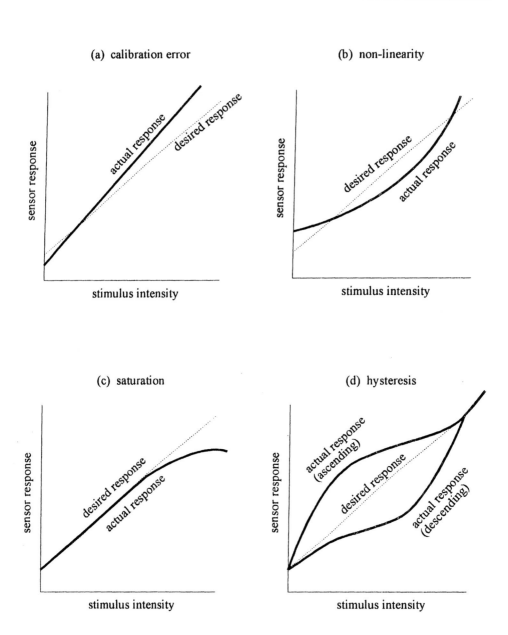

Figure 8-1. Varieties of Sensor Errors

practical for widespread use. This means that for each sensory mode there are only a limited choice of materials. It also explains why, for many possible sensory modes, artificial sensors of that type never leave the laboratory and why corresponding biological sense organs never occur (as discussed in Chapter 1).

The sensor designer can usually fix the sensor mode and field of view. Problems usually arise in the sensor band and the intensity range. Consider optical sensors, for example. There are only a limited number of materials that can act as photoreceptors to begin with. Among these, the intensity response over the entire spectrum of frequencies will not be constant. And the intensity range will not only be limited, it will likely be non-linear or non-logarithmic in different intensity regimes. All of this has to be corrected for or otherwise dealt with in the sensor design.

Material imperfections lead to errors of several varieties (Figure 8-1 and Reference 4):

1) *Calibration error* is due to the inability of the sensor designer and the sensor manufacturer to make the sensor output reliably represent the input sensor stimulus. For example, the translation of stimulus intensity into numerical output will never be exact: it can be in error both in the absolute level and in the slope of the response function. Calibration error thus introduces a systematic distortion into the resulting image.
2) *Non-linearity* error is specified for sensors which are supposed to translate the stimulus intensity linearly into a numerical output level. Due to the limitations of materials, perfect linearity is difficult to obtain over the entire intensity range.
3) Closely related to non-linearity error are the effects of *saturation*. A sensor material may respond to a stimulus in a linear fashion over a wide range of intensities until it nears an upper limit, the saturation point. At the saturation point, the response curve becomes highly non-linear, and beyond the saturation point, intensities are registered as if they were at saturation. Saturation is due to, once more, the limitations of materials. In a photoreceptor, for example, saturation reflects the fact that only a finite number of photons can be converted into an electrical signal within a given time. (Saturation occurs in human vision, for example, if we try to look at

something very bright, like the sun.)
4) At the other end of the scale are *threshold effects*. All actual sensors have an intensity threshold, a stimulus intensity value below which the sensor will not respond. This leads to two kinds of error. The first is the variation of the actual sensor threshold from what is expected (a form of calibration error). The second is that the sensor will continue to produce an output even if no stimulus is present. This is the phenomenon of the dark current, and is responsible for the "intrinsic light" of vision -- the fact that in a totally dark room we still perceive grayish patterns that come and go.

The above four errors are all static, in that they depend on the stimulus level and not changes in the stimulus. There are also, however, several dynamic errors stemming from material imperfections. These include:

5) *Hysteresis* results in an error in the sensor's output when a given point of an input signal is approached from the opposite direction. To use the example of Fraden[5], at 50°C, a thermometer shows 49° when the object is warmed up to that point and 51° when the object is cooled down from a higher temperature to 50°. This means that the thermometer has a hysteresis error of ±1°C.
6) Another dynamic source of errors is that of limited *repeatability*. Errors of this sort are due to the sensor's inability to represent the same value, under otherwise identical conditions, when the stimulus is repeated. Repeatability error may mean the decreased sensitivity of the sensor to the stimulus over time (a phenomenon akin to saturation), or it may mean the sensor keeps representing the same stimulus even after it is no long present. This latter dynamic error is familiar to us in visual afterimages.

8.4.2 ADDITIVE NOISE

The main source of error in the sensor itself is random noise. Noise has the effect of adding a random signal to the sensor image, producing random variations in intensity for each pixel. Some noise level is unavoidable, because the physical world is full of continual random fluctuations. We are all familiar with noisy images, although we may not call them by that name. The "snow" on the screen of a television turned to an unused channel or the uniform hiss of rushing air are examples of random noise. Listening closely to a radio, you will hear, behind the voice or music of interest, random (white) noise -- small relative to the signal for a nearby station and comparable to the signal itself for a distant station. What you hear is thus always a combination of signal and noise. Any pixel, then, when random noise is present will have the form: pixel output = signal + noise. Generally, the random error introduced by noise will introduce a Gaussian variation into the output, yielding an error of ±F and the probability of a data value being x:

$$P(x) = A e^{-(x-x_o)^2/\sigma^2}$$

where x_o is the true data value and A is a normalization factor.

There is also a random statistical error to take into account. The raw statistical error is inversely proportional to the square root of the number of events;
error = N^{-2}. In most uses of sensors, this is not a problem, due to the extremely large number of counts or physical quanta involved in the action of the object upon the sensor. For example, a photocell of 1 cm^2 area placed in direct sunlight would intercept approximately 10^{18} photons each second, leading to a statistical error of one part in a billion for such a sample. However, in the scientific use of sensors, event rates may be very low, making the raw statistical error the major source of uncertainty in the data.

8.4.3 SECONDARY MODES OF SENSATION

The primary mode of sensation is that for which the sensor was intended -- the primary mode of the eye is vision, the detection of light. But due to material limitations, sensors will also have secondary modes of sensation: the same sensor may be able to detect other types of energy in a limited way. If you press on the eye, it causes sensations of color. We can see sunlight, but we can also feel it as warmth on the skin -- a very rudimentary kind of sensing. Infrared light was actually discovered in this way: a thermometer responds to the invisible part of the solar spectrum. There are other examples. We can hear a very loud rumble like thunder or an explosion, but we can also feel it bodily. Intense ultrasonic sound waves are known to make people nauseous. We can hear a ringing bell, but also feel its vibrations by touching it. Secondary modes of sensation are a source of noise, and can actually interfere with the sensor's normal mode of operation. (Secondary modes of sensation should not be confused with the psychological phenomenon of synaesthesia, which is the confusion of sensory modes and qualities in the nervous system.)

8.5 Strategies for Dealing with Errors

Like the topic of errors and noise itself, methods of dealing with errors have received much attention. Many of these methods have emerged from the study of telecommunications and signal processing, fields with a keen interest in holding errors to a minimum. The interest of general sensor theory here is whether errors constitute an impenetrable roadblock between the sensor and the object, or whether they merely remove some portion of information (rather like projections). We will find that the latter is true: it is possible to extract the signal from the noise, although the signal will now be imperfect.

To repeat a point made earlier, if it is simply a question of projection, we may correct for it exactly if the transformation is known. It is, for example, possible to "deblur" photographs if the distance to the object, the lenses used, and so on, are known. Thus, projection in itself is not a source of error: it never adds extraneous information; it merely rearranges or reduces the information that is there in a well-understood way.

Truly random noise -- whether space, time, band, or intensity fluctuations -- can easily be removed, especially is the character of the desired signal is known. If not, it produces a constant background level that "submerges" the signal of interest. Because it is constant, it can be removed is the signal is not entirely drowned out by it. Systematic errors can also be easily removed if their nature is known. If not, they also form a background which can be filtered out. Generally, the influence of noise is to blur the signal or object of interest -- it reduces the information content of the sensory object. It decreases the effective resolution of the sensor, in other words. This degrades quality of the image, but it does not eliminate the information (now vaguer) about the object that the image contains.

There are several strategies beyond signal processing techniques that can be used to decrease errors and increase the all-important signal to noise ratio. First, one can simply move closer to the object, the most common human response to errors. Moving closer cuts out the amount of medium between the sensor and the object, and makes the object apparently bigger and more intense, which decreases the relative effects of errors in the sensor itself. Next, is to observe the object for a longer period of time, or to make several observations. (This assumes, of course, that the object is changing relatively slowly.) Or one can simply use a "bigger" sensor, with greater extent and resolution in band, field, and range. Another strategy is the use of two or more sensors to view the same object. This will greatly decrease the errors coming from the medium, as well as giving a more complete view of the object. Different perspectives decrease the probability of obscuration, as well as eliminating much random noise (random fluctuations in the medium are not spatially correlated). Extrapolating along the same lines, the use of several different modes of sensing can eliminate the errors stemming from a given mode. A multimode technique will be of advantage when the properties of interest are spatio-temporal. Not only are the effects of noise reduced, but also those of obscuration and attenuation (different for different stimuli), as well as inherent sensor errors. Finally, one can use an active sensor. Since the active sensor forms a feedback loop between itself and the object, it is often less subject to the influence of the medium. It can also actively probe the structure of the medium itself and provide information for the removal of its effects.

In a practical sense, all of these techniques are used by experimental science to obtain more extensive and more detailed information about the world, and to reduce ambiguities in that knowledge. The progress of science has been to a large degree one of "moving closer" to objects with ever more sophisticated sensing systems. Error can never be entirely eliminated; this fact does not, however, negate the possibility of knowing sensible objects.

8.6 SUMMARY: AN INFORMATION THEORY MODEL OF THE RELATION OF SENSOR AND OBJECT

Great strides have been made in recent decades in the understanding of communications in terms of information theory. Information theory is the study of how the laws of probability limit the design of information transmission systems[6]. Information theory models the communications channel between a transmitter and a receiver in terms of a digitized set of symbols. The number of symbols (e.g., 26 letters in the alphabet) are possible "states" of the transmission into which errors can be introduced. Information theory tries to understand how errors originate and how they can be dealt with.

A generalized communications system has five elements, as illustrated in the top half of Figure 8-2. A *source* produces the signal to be transmitted. The source might be a human voice for telephone or radio, a digital file for a computer network, or a video image for television. An encoder/transmitter translates the source signal into a digitized representation and sends it into the communications channel. The channel is the connection between the transmitter and the receiver; it can be a telephone wire, a fiber optic cable, a satellite downlink, a radio wave bounced from the ionosphere, and many other things. In the communications channel, noise will be introduced which degrades the desired signal, leading errors and data corruption. No actual communications link is completely noiseless. The signal with noise is then received and decoded, and the resulting decoded signal is sent to the user (normally a human being).

The relation of sensor and object is analogous to a communications system, as shown graphically in the bottom half of Figure 8-2. The source generates the distal stimulus. This can be a physical object or another information system for an intelligent object. The sensory object, the spatio-temporal field of interactions, corresponds to the encoder/transmitter. The medium is the same as the communications channel, and it is in the medium that noise is mixed with the stimulus. The role of the receiver/decoder is played by the sensor, as it captures and translates the stimulus. Finally, there is an exact correspondence between the user in a communications system and the information system of sensor theory.

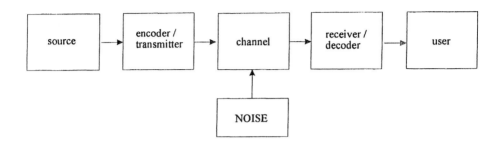

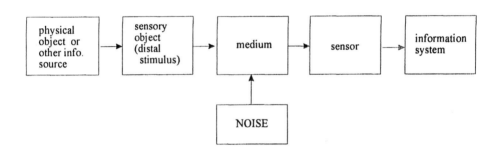

(b) Relation of Object and Sensor in Sensor Theory

Figure 8-2. The Relation of Sensor and Object as a Communications System.

The source of noise and errors will principally be the medium between the sensor and the object, just as in a communications channel. We have previously shown that the sensor data representation or image has the form of an array of information, and that its maximum information content can be measured. This, when combined with the sensor's effective frame rate determines the channel capacity (in bits/sec) of the sensor's relation to the object. For example, consider the screen graphics image discussed back in Chapter 3. It consists of 640 x 480 pixels of 24 bits each, a total of 7.37×10^6 bits/image. Suppose we have a sensor that produces such images at a frame rate of 30 per second. This implies a sensory channel capacity of 2.21 x 10^8 bits/sec (about 20 times the channel capacity of human vision).

The introduction of noise by the medium or the sensor reduces the effective channel capacity (it can, in fact, reduce it to zero). The error-degraded image introduces ambiguities into the possible knowledge of the object, and more errors, the more vague this information becomes. It is important to see that errors have this effect of blurring or smearing out one's information, rather than destroying it at a blow. It reduces the specificity of inferences that are possible from the information. Even when all particular information is lost, one can still infer that "something is out there" (even if that something is the entire world). The main effect of errors and noise is thus to decrease the resolution (intension) of the information regarding objects. Returning to the example in the previous paragraph, a large amount of noise with a high spatial frequency noise (e.g., "snow") could be introduced into a 640 x 480 pixel image without degrading information regarding the large scale shapes of things. This effect is seen easily, for example, by tuning into a distant television station: despite the noisiness and graininess of the image on the screen, which washes out fine details, larger shapes are still discerned.

The analogy of a communications system is relevant to sensor theory in another way. In all sensation, a representation of the properties of the object is being communicated in some fashion. The sensory process is thus the communication of the object and the information system through the intermediary steps of a stimulus and a sensor. Exactly what is communicated is the topic of the next chapter.

CHAPTER NINE

QUALITIES AND THEIR RELATION TO OBJECTS

9.1 THE PROBLEM OF QUALITIES

We now return to the question of qualities and their relation to various objects in the "real world". Do sense qualities objectively stand for properties of real things? Or, to the contrary, are they merely subjective and private sensations? And if there is a correspondence between qualities and objects, what is it? These are old and highly controverted questions of immense epistemological importance. The principal aim of this chapter is not to solve them once and for all, but to show the utility of sensor theory in understanding them. In particular, it expands the consideration of qualities beyond those of human perception and psychology to the most general sensor point of view.

By quality is meant, to repeat, a data representation perceived as belonging to an object. Chapter 3 dealt with this from the side of the sensor. In sensory transduction, the sensor translates a physical stimulus into a data representation. The data representation or image which the sensor outputs is a spatio-temporal array with intensity and qualities for each pixel. These representations may be numerical or symbolic. We saw that all sensation has an intensity: a magnitude or quantity of sensation, representing the total energy of the stimulus. We perceive intensities as brightness, loudness, and so on, depending on the sensor mode. There are two types of representations denoted qualities as such. The simple qualities have no reference to space or time (i.e., are unextended and "pointlike") and represent the band parameters of the stimulus. These simple qualities are essentially submodalities or qualifiers of the intensity: they mean intensity of a certain kind. They are always mode-dependent. We perceive them as colors, tones, scents, tastes, heat, cold, and so on. There are also, we saw, complex qualities. These do have a finite extension in space and/or time -- they are essentially spatial or temporal patterns, such as shapes and textures. When they represent patterns of intensities alone, they are mode-independent.

Yet the description of data representations given in Chapter 3 does not deal with every aspect of quality. What that description gave us were the representations which correspond to the quality from the side of the sensor -- that is, sensation. But the quality is the representation *as it is perceived*. The question is, in qualities, do we perceive things and their properties (however indirectly) or do we perceive something purely subjective, belonging to the mind? In other words, what do sensory representations represent? The validity of sense knowledge thus turns upon the validity of intensities and qualities, for they are the sensory point of contact between the knower and the material object known.

The crucial question is thus whether the qualities stand for real properties of the objects. If qualities are real and objective, then they *must correspond* to something in the "real world". Even if they are subjective representations, they can only exist in mind according to the structure of the object; they are not arbitrary. Sensations would thus be *symbols*, having no meaning apart from the objects to which they refer. Conversely, is it possible, even in theory, to have a sensation that is *totally* different from its source -- one that bears no fingerprints of its maker, so to speak?

Common sense assumes qualities are real and objective. So did the Aristotelian-Scholastic philosophy that governed Western thought for so many centuries. But modern philosophy began to chip away at this view, leading to the skeptical denial of any firm correspondence between the object and qualities, or, in contrast, the idealistic identification of reality with the qualities, denying any object beyond them. Views on the status of sense qualities thus form a dividing line between philosophic schools.

We have already seen some of the difficulties that arise in a consideration of human sense perception. Human beings are not equipped with Argus sensors. Thus, any human view of objects is always from some limited perspective. We have access through our unaided senses to only a small portion of the sensory object. So even if the Argus sensor (and its qualities) are reliable, what of the much more limited qualities of human perception? At the same time, the previous chapter's discussion of perspective and projection removes several problems for us. Projection alone does not eliminate the possibility of valid sensory knowledge; it merely restricts it. It yields partial information. The relation between the sensor and the object can be understood in terms of information theory as a sort of communication. The physical properties of an object give rise to a distinctive spectrum of interactions -- there is a consistent mapping between the physical object and the sensory object. This sensory object is then projected onto the sensor. It was also shown in Chapter 2 that the protosensor must always be reliable in its representation of a stimulus. All this speaks in favor of the validity of sense qualities. What remains is to take the final step into the information system itself.

To fully understand the issues involved here, the chapter will first briefly review the theories of sense qualities as they have developed. Then, it will discuss in turn the relation of intensity, simple qualities, and complex qualities to objects, from the angles of both sensor theory and perception. Finally, the important questions of intentionality and mutual conformation of sensory information and objective properties will be discussed.

9.2 THE ARISTOTELIAN THEORY OF SENSE QUALITIES

Speculation on the nature of sensation is doubtless as old as philosophy. We know but little of these early theories. The Epicureans, for example, held that in sensation a little image (*eidola*) of the object penetrates the sense organs, and that sense perception is always true. For Empedocles, sensation occurs because things are always giving off effluence, which enter properly-sized pores in the sense organs. Plato almost entirely passed by the subject, because he asserted that sense perception is relative and thus not knowledge.

The earliest comprehensive scientific and philosophical treatment of sensation was provided by Aristotle. This should not surprise us, as Aristotle, in contrast to Plato, based his entire theory of knowledge on the possibility of true knowledge of the world of the senses. And as Aristotle's theory of sensation was taken up almost unchanged by later Arabian and Western Scholastic thinkers, it formed the starting point for all discussion of sensation in Western thought.

Aristotle held that sensation is an essential power of the animal "soul", and is directly linked with complementary powers of desire and motion. In other words, animals have senses because they must actively seek or avoid things in the material world.

The sensory process for Aristotle begins with existing material objects, particular beings composed of matter and form. It is these particularized forms that announce themselves in sensation. And I think "announce" is the right word, for it is as if everything in the world is calling out its name, whether clearly or obscurely. The qualities (accidental forms) of things, such as color or shape, are conveyed to the observer through a medium and then sensed. Aristotle supposed, for example, that colors exist on the surfaces of bodies, and the colors are transported to the observer by setting the air in motion[1]. The form of the color is precisely the same in the eye as in the object. And likewise for all the other sense qualities: the form of the object is impressed upon the sense organ. Here it is worth quoting Aristotle's famous "signet ring" analogy:

By a 'sense' is meant what has the power of receiving into itself the sensible forms of things without the matter. This must be conceived of as taking place in the way in which a piece of wax takes on the impress of a signet ring without the iron or gold ... in a similar way the sense is affected by what is colored or flavored or sounding, but it is indifferent what in each case the substance is, what alone matters is what quality it has.[2]

These impressed forms are the *sensible species*[3]. The qualitative change made by the sensible species in the sense organs is essential to the cognitive process, for the sensory powers are entirely passive. Aristotle draws an analogy between sense and the combustible -- something from outside is required to set it going. The sense organ not only needs to be moved, it also needs to be determined or specified to a given quality: to see red rather than blue, for example. (There is no generic sensation, only specific -- hence the term sensible *species*.)

For Aristotle, the sensible species are the principle of all human knowledge. However high our minds ascend, they must begin with the quiddities of material beings. The sensible species are what make the contact of mind, senses, and *particular* things possible. (There are also intelligible species, or concepts, through which the mind grasps the *universal* natures of things.) They are a bridge between the knower and the known. The species itself is not the object of knowledge. It is, rather, the medium by which the real object and its qualities are known. It is the representation of the object *to* (and *in*) the knowing subject, an actual connection of knower and known. The species can play this role of bridge precisely because they are the form or similitude of the thing known. It is a more or less complete, but entirely accurate, likeness of the thing. The form of the thing known is actually conveyed to the knower, and becomes immanent in the mind. The species connects the knower with the very nature of the known through its form. This also means that the sensible species have intentionality: they exist in the knower only with reference to the object known. We will return to the question of intentionality below in section 9.5.

The sensible species are proportioned to the sense mode that receives them -- that is, our five external senses of sight, hearing, touch, taste, and smell. Each sense has its proper sensibles. The proper sensible and object of vision, for example, is color. For Aristotle, a rose or a ruby or a neon sign continually radiates particular forms of red, which we apprehend. The proper sensibles are grouped into genera by sense, and there are only a finite number of sensible species possible for each sense. Aristotle argued that the species of all sensible qualities are limited, because for all things between extremes, there can only be a finite number, and sensible qualities are like this[4]. And our perception of these proper sensibles is never in error, or admits the least falsehood of all our knowledge. The proper sensibles are what Aristotle calls affective qualities, and correspond to the later secondary qualities, and to the simple qualities of the present work. There are also common sensibles -- the sensible species of motion, rest, figure, extension, number, and unity[5]. Note the prominence of mathematical properties here -- they correspond to the later primary qualities or complex qualities. Perception of the common sensibles themselves requires no special faculty, as they are found in each sense. Rather, they are what permit us to behold the unity of the object with several different senses.

Thus, we see that for Aristotle, qualities are always objective. They are real properties of objects, and this is the case for both the common sensibles like figure and extension and the proper sensibles like color or taste. Such qualities, conveyed to the knower, put the knower in contact with the form of the object -- its true nature -- a contact which grows in sophistication with additional information from each sense.

9.3 MODERN THOUGHT ON QUALITIES

The scientific revolution of the 17th Century brought the downfall not only of Aristotle's physics and cosmology, but also his theory of sensation. Perhaps fittingly, Galileo was probably the first to draw the distinction between primary and secondary qualities[6]. Primary qualities -- mathematical in character and roughly the same as the old common sensibles -- were the only ones that mattered scientifically. As Galileo wrote in a famous passage in *The Assayer*, the language of nature is mathematics, and its characters are triangles, circles, and other geometrical figures. The secondary qualities -- proper sensibles like color -- were demoted to an inferior level. They do not really belong to the object, and yield no true knowledge. Secondary qualities are not properties of objects. They are subjective and dependent on the sense organs.

This step taken by Galileo -- the father of modern science -- was driven to its logical conclusion by Descartes -- the father of modern philosophy. In the world of mathematical bodies, sensible qualities as such have no role to play. The objective and intentional sensible species were replaced by subjective sensations, and this notion became an almost universal philosophic assumption. In other words, when we perceive a color, it is in us, and we are perceiving a sensation, not the real property of a thing. Bodies themselves are purely mathematical, characterized by figure and extension. Secondary qualities are nowhere in the world, but in the mind. Sensations are related to outer things not intentionally, but as effects to causes. Sensations as such have no cognitive utility. It is this theory of sensations that allows the Cartesian dream argument and its denial of a real difference between sensing and imagining. For Descartes, the external world of bodies can be reached only through mathematics and the guarantee of God. In other words, the sensory bridge between the knower and material being collapsed, except for one last mathematical cable spanning the chasm.

The general trend of this argument was taken up later by Locke and the other empiricists, who made two main assumptions. First, precisely because sensations are relative, they are also subjective. We are all familiar with the relative character of sensible qualities. Yet the relativity of the secondary qualities was taken as direct evidence of their subjectivity as sense-impressions[7]. The second empiricist assumption was that we immediately perceive only our sensations or ideas -- the "ideas of sensation and reflection" in Locke. Yet Locke still thought we could know at least the primary qualities of things. Berkeley disposed of that assumption by showing that our perception of primary qualities depends on the secondary ones -- without color, for example, we could not see the figure and extension of objects. Thus, the sensible world entirely consists of congeries of subjective sensations and there are no material things. Behind the sense qualities, there is nothing but our own minds and a divinely-supplied order. Descartes' mathematical cable to the material world snapped. From there it was but a small step to Hume's skepticism, which completed the empiricist course by getting rid of the self as well. All we are left with is a perceptual stream of "impressions" and "ideas".

On reflection, it is relatively easy to see how the common-sense view of qualities can be put into doubt. The sensory appearances of things -- at least their appearance to the human senses -- do not allow us to grasp their inner natures. I see a shiny yellowish metal object in front of me. Is it made of gold or brass? If it is gold, does its sensory appearance of yellow, shiny, heavy, and solid allow me to deduce that gold is composed of atoms with 79 electrons orbiting a nucleus of 79 protons and 118 neutrons? Of course not. And it is a fact, as was discussed in the previous chapter, that our perceptions of things are relative and limited. Thus it seems that qualities -- at least the simple qualities -- do not correspond to the object at all. When I see that gold is yellow or feel that it is heavy, these are just perceptions of my own representations or sensations. In opposition to uninformed common sense, the qualities now seem to belong entirely to me, not the object. When we perceive qualities, we know them only; they do not yield knowledge about material objects:

When I see a tomato there is much that I can doubt. I can doubt whether it is a tomato I am seeing, and not a cleverly painted piece of wax. I can doubt whether there is any material thing there at all ... One thing, however, I cannot doubt: that there exists a red patch of a round and somewhat bulgy shape, standing out from a background of other color patches, and having a certain visual depth, and that the whole field of color is directly present to my consciousness.[8]

Thus, the whole question of qualities is really the question of the status of the data representations that are perceived at the end of the sensory process. If they must correspond in some way to the properties of the object, then a realist view of sensation and sense-knowledge is possible: even if rejecting some naive assumptions of common sense about the true natures of things. In sensation, we would perceive real things. If not, then we perceive only subjective sense data or quales or impressions with no firm relation to objects.

In the past two centuries, epistemological positions on sense qualities have solidified around those two possibilities. The subjectivist camp, following onto the aforementioned tradition of Berkeley and Hume, holds that qualities are of purely subjective importance and do not yield knowledge about things. In sensation, we are aware not of things, but of subjective sensations. This position quite naturally leads either to some kind of idealism or to an empirical skepticism, whether it be Mill's phenomenalism or Ayer's logical positivism[9]. For if sensations are always present as a subjective mental screen between us and the world, it follows we can know nothing certain about the world.

Even when a thinker desires to preserve something real behind the sensations, the logic of subjectivity expresses itself. The example of Kant and the later course of German idealism is instructive. When Kant took up the reconstruction of philosophy from the damage done to it by Hume's skepticism, he still held nonetheless to the general notion of sense-impressions. At the same time, he wished to preserve real things-in-themselves behind the sensible appearances. The sensations, caused in us by the things-in-themselves, in no way resemble their cause. Not only do secondary qualities tell us nothing about the things-in-themselves, neither do the primary qualities: the things-in-themselves are not in space and time. But then it was easy enough to then suppress the real thing-in-itself as unnecessary (and as improperly using the category of causation), yielding a straight idealism. And that is exactly what the later idealists did. Even the phenomenological school, which has done such an admirable job of restoring meaning to sense qualities, has not broken with the subjectivist view of sensation[10].

There are also those who still hold to an objectivist view of sense qualities. For them, as for Aristotle, qualities have a real meaning and correspondence to the properties of objects. In sense-perception, what we perceive are objective things, not subjective sense-data or deputies. The objectivist camp is one of realists. On the one hand, it contains neo-Aristotelians and neo-Thomists, who return to the Aristotelian view of sensation as the conveyance of the form of a material thing. On the other hand, there are the Neo-Realist followers of Thomas Reid and the Scottish School of Common Sense[11]. The cornerstone of Reid's theory is the rejection of the notion of Descartes, Locke, Berkeley, and Hume that we perceive sensations rather than things. Common sense supposes that we perceive things, and Reid held that common sense is always right. Reid's theory of sensation had a large influence on the 20th Century school of Direct Realism[12].

One point, at least, is clear: the objectivity of qualities -- at least of the primary qualities -- is essential to any realist epistemology. Without that objectivity, one ends up with idealism or skepticism. The objectivity versus subjectivity of sensation and sense qualities is thus the boundary line between entirely different classes of epistemologies. And we will now see that general sensor theory lends support to many aspects of the realist view.

9.4 THE OBJECTIVITY OF QUALITIES IN GENERAL SENSOR THEORY

Chapter 3 of this book set out a new division of qualities. In all sensation, there is an intensity, and intensities can underpin two kinds of quality. There are the simple qualities, which require no extension in space or time. They depend only on the parameters of the physical carrier and the sensor, and they can be grasped by a single sensor element. The simple qualities correspond roughly to the old secondary qualities, such as color. Each is characteristic of a single sensor mode, and may have an inherent ordering (like the color circle). In addition to the simple qualities, there are complex qualities, which can only be perceived as extended in space and/or time. They depend upon spatial or temporal patterns of intensities or simple qualities. If the extension is not directly perceived in the quality, then one has a microcomplex quality, such as a shiny or dull surface. If the extension is directly perceived, as in the shape of an object, then one has a macrocomplex quality. Macrocomplex qualities correspond to the old primary qualities, and like them, they may be perceived by more than one mode of sensor (which is why they can provide a basis for multimode sensor fusion).

As discussed previously, this section examines what sensor theory has to say regarding the objectivity of simple and complex qualities, expanding upon the discussion in Chapter 3. Is the nature of the quality determined by the object or the subject? If qualities are objective, then they stand for the properties of things outside the subject, whether directly or indirectly. If qualities are a reliable and consistent mapping of the properties of the objects, then they can be a source of knowledge about those objects. They would be like a sign pointing to the properties of the object. Conversely, as was seen above, the subjectivity of sensation renders sensible knowledge of real things impossible. The aim here is to examine what sensor theory implies about the relation between qualities and objects. The next section (9.5) will show why this is the only explanation that works.

In brief, sensor theory implies that qualities are, at least in part, objectively valid. Qualities, however partially or vaguely, correspond to the properties of objects. They are the final stage of the projection of the object as perceived. This does not rule out a subjective element in sensation -- for simple qualities, this subjective element is important. Qualities, to repeat, are "forms of sensation". In transduction, the organizing principle in the stimulus is transposed onto something within the information system, whether it be an electrical current, digitized numbers, or a train of pulses. We have also seen from the discussion on projection that sensible knowledge admits of degrees: the partial nature of our sense-perception, our limitations, do not cancel out the validity or objectivity of qualities. The real question is, once more, is whether it is possible to have a caused sensation whose properties are completely different from its source. Sensor theory's answer to this question is, no -- there must be at least some correspondence between the object and the quality for sensation to occur at all. Qualities are, in essence, the properties of the object as projected (or transmitted) through stimuli, reduced to the sensor conditions, and then perceived.

The partial knowledge yielded by sense-qualities takes the following form. I perceive a quality X -- say, redness or bitterness. X defines a class of possibly existing objects. When I perceive X, then I can reliably assert that at least one member of the class of objects must exist and be present. The more qualities perceived, the more detailed and specific is my sensory knowledge. The number of perceivable qualities depends upon the extension and resolution, so that for the Argus sensor, knowledge of sensory objects would be complete and perfect.

In other words, with increasing resolution, perception of the properties of things will *converge to a limit*. Convergence is ultimately the measure of the truth of sensation. If convergence does not occur (and for intelligent emitters, it need not), then sensation cannot give us complete knowledge of the thing sensed. To the degree there is convergence, however, sensation is a source of knowledge.

The Argus sensor also reminds us that we are dealing with sense-qualities *in general*, not just with those qualities perceivable by human beings. To demonstrate the failings of human sensation under this or that specific condition does not negate the validity of sensation in general: the previous chapter showed, for example, there are no general exterior illusions. We must go beyond human sensation to the general point of view; active sensing, for example, provides a powerful tool in understanding the nature of qualities.

The question of the objectivity of qualities naturally divides itself into (1) intensities, (2) complex qualities, and (3) simple qualities. We will deal first with intensity, because both simple and complex qualities ultimately depend on intensity. The simple qualities are specifiers of intensity, while the complex qualities are intensity patterns. Next, the complex qualities will be dealt with, because the nature of their objectivity is more obvious than for the simple qualities, which will be treated last.

9.4.1 INTENSITY

Let us first review what the earlier chapters have told us about intensity.

Intensity is the most fundamental aspect of sensation and is inseparable from it. All sensation has an intensity, because there must be an energy transfer to set the sensory process in motion. We perceive intensities as brightness of light, loudness of sound, and so on. One can have intensity without any qualities, but not the reverse. There is a single modal intensity for each pixel, but several possible qualities. Intensity stands for the total energy or number flux involved in the sensation. It is thus the quantity or magnitude of the sensation, and for this reason is the most quantifiable of aspect of sensation.

The objectivity of intensity is fundamental, for the simple and complex qualities ultimately depend on it. The objects we see have shapes, but these shapes always have varying brightnesses which permit us to locate them. The words we hear have a certain combination of loudnesses. The edges and surfaces we feel would be imperceptible without hardness.

We have seen that in the case of animals and humans that there are well-defined laws relating the intensity of the stimulus to a frequency of nerve impulses. In other words, there is a mapping between the two. The stimulus intensity -- which means the degree of interaction or nearness of the object and the sensor -- is thus dependent on the relation of the object and the sensor. The internal encoding, the data representation, of the intensity may be subjective (i.e., peculiar to that information system), but the mapping of the stimulus into it is an objective (and *a priori*) rule. Intensities as perceived are thus objective -- they are a projection of the magnitude of the stimulus.

Intensity stands for the degree of connection of the observer and the object, the effective "nearness" or "farness" of the two, because the intensity stands for the energy involved in the relation. It is thus characteristic of the relation itself: at zero intensity, there is no sensible relation of the object and the sensor. Perceptible intensity, of its very nature, presupposes an object, a sensor, and the relation of the two through a stimulus energy.

This may be seen in another way from active sensing. An active sensor beams out energy and then receives back its own energy after it has interacted with an object. It is obvious that the energy intensity received back by the active sensor from the object (reflected or transmitted) must be less than or equal to the energy intensity sent out. The energy beamed out by the active sensor will also have a spatio-temporal pattern of intensities. One sees immediately that the return signal, as reflected or transmitted by the object, must be some subset of this pattern. As the distance between the active sensor and the object decreases, the intensity of the return signal increases or remains constant. The time lag between the incident and return signal will decrease with decreasing distance. On the other hand, as the distance between the active sensor and object go to infinity, the intensity will either decrease to the vanishing point or at best remain constant. The time lag will go to infinity with the distance (meaning that an infinitely remote object could never be sensed in finite time). All this shows that intensity is indeed the degree of connection, the nearness or farness, of the sensor and the object. Except for a perfectly collimated beam of energy, intensities must decrease with distance. This shows us as well the essentially spatial character of sensible intensities. And it is doubtless from active sensing -- touching and moving objects -- that our common sense certainty of the existence of the world arises.

The simplest inference from a perceived intensity is just like that from a protosensor: something with the sensor conditions is present, and a physical relation exists with it, however tenuous that relation might be. This thing is not inside the information system, because it requires the intermediary of a sensor to be perceived. A further inference would be that the sensible thing perceived is acting strongly or weakly upon the observer.

9.4.2 COMPLEX QUALITIES

To reiterate, it is best to deal with complex qualities before simple qualities, because complex qualities are essentially spatio-temporal patterns of intensities, and their objectivity is thus more direct and obvious than that of simple qualities. Once the objectivity of intensity is understood, the objectivity of complex qualities follows. Chapter 3 showed that the complex qualities are themselves divided into macrocomplex and microcomplex. In macrocomplex qualities, the spatial or temporal extent of the pattern is perceived (and is usually the information sought), while for microcomplex qualities, the extent is not directly perceived. The distinction of the two depends on the capabilities and limitations of the sensor and information system involved.

The macrocomplex (primary) qualities are a direct projection of the spatial or temporal structure of the object. We directly perceive these qualities (such as shape) as they actually belong to the projected sensory object. This follows from the discussion in the previous chapters. The sensory object has spatial extent, temporal duration, and displays various intensity patterns in both. The mapping of the sensory object into the sensor image is just a projection, and follows objective geometric laws. The object is a spatio-temporal field of intensities; the complex qualities are an array of quantized intensities as perceived. Such a mapping is *relative*, but it is not *subjective*. (What subjectivity is found in macrocomplex qualities is due to the transduction of intensities.) It is entirely reliable in theory. The complex sensor can detect extended patterns because the sensor itself is extended. The sensor image is a field of potential patterns which are actualized by the stimulus pattern -- there are as many possible macrocomplex qualities as patterns. Complex qualities depend on intensity variations.

The spatial or temporal pattern of an object is quite literally impressed on the sense organs. Our eyes, for example, focus a real (although inverted) image of things on the retina. If we see a house, the image of the house is present at the back of the eye, just as a camera places an image on film. This spatial pattern on the retina is conveyed by the optic nerve in an isomorphic fashion. Spatial information is most definite and complete in sight and touch, and is very carefully processed by the brain. We perceive these qualities with more than one sense -- shape can be known by both vision and touch, and blind mathematicians are possible. Experiments have shown that spatial information is shared between sensory modes in the nervous system.

The sophistication of the grasp of complex qualities depends, however, on the spatial or temporal resolution of the sensor. (This is another indicator of the essentially isomorphic grasp of complex qualities.) The spatial discrimination of vision is excellent, because there are millions of rods and cones spatially arrayed in the retina. Touch (especially in the fingertips) also has a great spatial density of receptors. But the spatiality of hearing is much less acute than vision, while that of smell or taste is very weak indeed. Thus, our knowledge of the complex qualities will be sharp or vague, depending on the sensor involved. When the spatial resolution of the sensor is poor, rebinning occurs, and finer complex qualities cannot be perceived. Detail is blurred. We are all familiar with this in, for example, the recognition of someone at a distance. At first, recognition may be impossible, because of a lack of detail; as we move closer, recognition becomes possible with more detail -- that is, better grasp of the complex qualities. This shows us quite directly the objectivity of macrocomplex qualities by the measure of convergence. With increasing sensor extent and resolution, the perceived macrocomplex qualities (spatio-temporal properties) of an object converge to a limit. The Argus sensor would yield a perfect image of the shape and so on of the object. This convergence with sensor improvement could only occur if the complex qualities really stand for the properties of the object.

All complex qualities, to repeat, have some spatial or temporal extent. They could not be perceived otherwise. (Our perception of complex qualities proves that we can perceive a spatial or temporal field of intensities as a single whole -- we perceive objects as objects, not congeries of sensations.) Complex qualities represent the "internal relations" of the object; that is, the spatial or temporal arrangement of the parts of the object. It is the "picture" of these relations which is conveyed to us and projected upon the sense organ or sensor. The complex qualities thus do correspond to the object: the spatio-temporal structure of the image is a projection of the spatio-temporal structure of the object. The image will generally be an incomplete picture of the object, of course, but what it does convey to us is entirely reliable. It is an objective resemblance.

This is confirmed by the fact that the main purpose of our senses with high spatial resolution, such as vision and touch, is precisely to gather spatial information about objects: their locations, shapes, sizes, and motions. Human sensation does this imperfectly, of course. But it does convey enough reliable spatial information about the world to allow us to move about in the world and live our everyday lives. It is imperfect, but not useless.

Microcomplex qualities are also based on spatio-temporal patterns, but their spatial or temporal nature is not directly perceived. They are generally based, as was discussed in Chapter 3, on a repetitive or quasi-random pattern. Consider a common microcomplex quality, texture. Velvet, for example, has a distinctive "velvety" look and feel. This quality is based on the structure of the object on a microscopic level -- there is a spatial array of short threads. We do not perceive this structure directly; instead, we have the impression of a velvety surface. This fact of human perception has been used to great effect by painters to make us perceive textures in an image that are not "really" there. However, we may confirm the "real" nature of the microcomplex qualities by a closer examination of the object. Velvet seen in a magnifying glass no longer looks velvety; it looks fibrous. In other words, the microcomplex qualities pass over into macrocomplex ones at a different resolution. It depends on the sensor and information system involved. Microcomplex qualities are telling us something indirectly regarding the spatial or temporal structure of the object, while the macrocomplex qualities are direct. The microcomplex quality is a shorthand or sign of an underlying spatial structure. An additional mapping or translation takes place in comparison to the macrocomplex qualities, and yet, as we have seen, this mapping is objective and reliable. The exact perception of a texture may be subjective; its relation to the object is not. It stands for something real in the object.

The symbolic nature of microcomplex qualities show us how to make the transition to the objectivity of the simple qualities. The exact way in which microcomplex qualities are perceived is indeed subjective. But what microcomplex qualities signify is objective. The symbolization may be subjective, but it will never be actualized without its corresponding pattern stimulus, and is meaningless except in reference to it.

9.4.3 SIMPLE QUALITIES

The objective meanings of intensity and complex qualities are fairly obvious and easy to understand. It seems entirely reasonable, for example, that shape truly belongs to the thing we perceive it belonging to. When, however, we consider the simple qualities -- colors, tones, smells, tastes -- it is much less obvious and much more controversial what relation, if any, they have to the properties of the object. The Aristotelian theory held that colors and the other simple qualities really are in the object. A red object really has redness, and it transmits red forms to an observer. But modern philosophy began with the assertion that such secondary or affective qualities are not in the object, but in the observer. They are subjective sensations, rather than objective properties. A sugar cube in itself does not have properties of whiteness or sweetness. These are merely subjective perceptions on our part: secondary qualities are not in the thing, but in our minds. Being subjective, they tells us nothing real about the object, and are thus epistemologically worthless. They are even deceptive, because they incline us to believe we know something about the object through them.

And it is a fact, that in the case of the senses of humans and animals, the purpose of the simple qualities is reliable identification in a natural context, rather than the scientific grasp of the properties of things. Fortunately for us, the natural world is "lumpy"; it is divided up into classes of objects that can be identified from relatively few features. Thus, it should not surprise us that the simple qualities do not tell us very much about things. They give us only that information which is biologically useful.

It is certain that simple qualities are not properties of objects as such. Objects that appear red do not possess "redness". It is also certain that there is a subjectivity in simple qualities. Perception of them varies from observer to observer. Thus, they have some subjective component -- they are not direct projections of the properties of the object in the way that complex qualities are. Does this make them purely subjective and meaningless, then? By no means. The relation of the simple qualities to the properties of the object is *indirect*, and usually quite vague. But vagueness does not cancel out knowledge -- it just makes it partial. In other words, for the simple qualities, there is an additional mapping from the interaction with the object to what we consciously perceive or the data representation in an information system. Here, unlike for the complex qualities, what we perceive are indeed "sensations". The question is whether these sensations objectively stand for something. Are they primitive signs of the properties of the object, however indistinct? The reliability of the simple qualities is the reliability of the mapping of the interactions into the sensations. If it is reliable, then the nature of the sensations are specified and determined by the nature of the object.

We can return here to the analogy with microcomplex qualities. These are essentially symbolic, but correspond to something real in the object. This is also the nature of the simple qualities. In Chapter 3, we saw that the data representations perceived as simple qualities are symbolic or encoded representations of the band parameters of the physical interaction. Colors represent the frequency of light, pitches stand for the frequency of sound, smells and tastes correspond to various chemical substances. The problems arise from the *vagueness* of these qualities -- the band parameters are rebinned to just a few simple qualities. All chemical substances, for example, are mapped into just four primary flavors. This means that, for any of our senses, we will perceive very different substances as having the same simple qualities. Many different chemical substances may appear the same shade of blue; both sugar and saccharin taste sweet. Likewise, the same chemical substance can assume different qualitative forms -- H_2O can appear to us as warm liquid water or as cold solid ice.

Yet vagueness, to repeat, does not cancel out knowledge. To our vision, for example, the glow of a neon sign appears reddish. We could cover a white light with a red filter, and it would appear the same to us. Indeed, there may be any number of physical causes that stimulate a perception of the same color: incandescence, electronic transitions, molecular vibrations, refraction, scattering, etc.[13] Yet if we view the neon sign with a spectrometer, a very different picture emerges. We would see not a single color, but a spectrum of atomic lines. This spectrum is unique; it is the "signature" of the neon atom, and no other spectrum can be exactly like it. Its structure corresponds to the internal structure of the atom, as the atomic lines show us the energy states of the atom. Not only do the "secondary qualities" in this case correspond to the atomic structure, they were the way in which that structure was discovered. Our own visual perception of the glow is indistinct, but it is not incorrect or meaningless. Grasp of the nature of the object improves with the sophistication of the sensor -- there is convergence. The Argus sensor could make positive identification of everything there is, and provide materials for the grasp of their natures.

We have seen that sensation will be possible only if the object interacts with some physical vehicle, but each physical object will interact in a unique way, corresponding to its nature. This means that if we could look at the entire spectrum of physical interactions with something, we would obtain its sensory signature. This "signature" is the mapping or projection of the form of the object. Whenever we sense something, we are thus receiving -- whether clearly or confusedly, wholly or partially -- the projected properties of the object. The simple qualities thus correspond -- however vaguely -- to something in the object. My perception of redness, for example, defines or delimits a class of possible objects, just as a two-dimensional projection defines a class of possible three-dimensional objects. Simple qualities are not "pictures" of the properties of the object, but symbols of them. Qualities may quantized, but the resulting numbers have meaning only as symbols.

But if the complex qualities represent the relations present within objects, the simple qualities -- like intensities -- represent a kind of relation of the object to the observer. They are an indicator that some relation of the object and the sensor exists. They correspond to the object, but are not "in" it as such; they are characteristic of the relation.

The nature of the objectivity can be better understood by recourse to active sensing. Consider a wine glass. If I tap the side of the glass, it will ring with a musical pitch. Now, suppose I have an active audial sensor -- a combination of a speaker and a microphone -- that beams sound waves of exactly that pitch at the glass. The glass will begin to resonate. (The glass can in fact be made to resonate to the point it shatters, that favorite stunt of opera singers). I can discover that resonant pitch by varying the frequency of the sound beam. The same simple quality, if turned into a command for an actuator, excites the object. One can have feedback of the musical tone into the microphone and back out through a speaker (a phenomenon familiar to anyone who has dealt with public address systems). Thus, the pitch I perceive from the wine glass if tapped does correspond to something real in the object: the resonant frequency is determined by the density and flexibility of the glass, its exact shape, and so on.

A subjective side of sensations remains, of course. What is subjective is the set of symbols chosen for the simple qualities for a given sensory mode. The representation depends on the sensor and information system in question. The representation is not, however, entirely arbitrary. It has a form or order. Colors are arranged in a circle; while pitches form an ascending series. This ordering is precisely the mapping function, and it is an *a priori* form. My perception of a given color may be subjective, in that it is different from your perception. There may even be variations within the same subject: my right and left eyes sense redness slightly differently, for example. Nonetheless, the *ordering* of colors is the same for everyone. Orange always comes between red and yellow on the color wheel; gray always forms its neutral center. (Gray, white, black, and the purity of qualities apply to all simple qualities.) The order is objective, and corresponds in some way to the order of the parameter space of the physical interaction. In the spectrum of visual light, the lowest frequencies of light are red, and form an ascending series through orange, yellow, green, blue, up to the highest frequency, violet. Our trichromatic visual system maps this frequency spectrum into a circle. The sequence is the same, except it jumps from violet to red.

In summary, then, the simple qualities are subjectively chosen signs of something objective. We may consider them "natural primitive signs" or even "primitive concepts" if we like. They provide a point of contact between the bundle of physical interactions with the object, however obscure it is. At the same time, however, we cannot directly grasp the physical nature of the object from its simple qualities. What simple qualities provide us with is a means of identification.

9.5 QUALITIES AND PERCEPTION

Sensor theory implies the objectivity of qualities, however remote the relation might be to the properties of the object. Now we will see why this is the only valid explanation of sense qualities. The key point here is that if sense perception is as sensor theory says it is, it necessarily involves the contact, in one way or another, of the mind of the knower with the nature of the known. This means that qualities must put us in contact with the object's properties. It would be impossible to have sensations at all without some correspondence of the sensation with the nature of the object. Some information about the object *must* be conveyed to the knower; there could be no "empty" causing of sensations[14]. In sense-perception, we would be perceiving the object. Conversely, if sense-data has no necessary connection at all with the properties of the object (i.e., an "empty" causing of sensation is possible), then we perceive only that data, not the object.

If sensor theory is correct, then qualities would be objective, in that they stand for something in the object. They would also, however, be *real*, in that the properties of the object are not necessarily exhausted by the qualities of human sense-perception[15]. In other words, the sensory object is not a product of my personal consciousness; it is real and independent of me. (Conversely, products of the imagination are completely exhausted by our sense-qualities.) Qualities are thus an overlap -- partial, not complete -- of mind with the object perceived.

9.5.1 QUALITIES FROM THE PERCEIVER OUTWARDS: THE IMAGE IS THE INTERFACE

The foregoing discussion has assumed the notion, shared by common sense and natural science, that the objects we perceive really exist separately from us -- that the world is real. Qualities have been described *from the outside in*. But this is still inadequate from an epistemological point of view. What remains is to explore qualities *from the inside out*: given the fact that we perceive qualities, how can we determine their meaning? The introduction of the conscious perceiver is what leads to most of the philosophical problems regarding sense qualities. For how does (or can) the perceiver know there is anything behind data representations?

In other words, we must return to the discussion of perception in the first chapter. What do we directly perceive? It is a world, apparently exterior to us, of spatially-extended regions, that both change and endure in time, with intensities and qualities. Without intensities and qualities, there is no sense perception. Qualities are the contents of sense perception. Sensor theory can tell us what representation corresponds to red, but it cannot tell us why red is red. Redness as such must be perceived to be understood.

Left at this point, we would be led to a purely phenomenalistic account of the world akin to Berkeley's. Things *really are* the qualities. They are reducible to sensations (or "possible sensations") and there is nothing "behind" them.

But such an account of sense perception leaves out some essential points. First of all, when we sense, it is always through a sensor, located on the body, as given in the image of the body. Likewise, we can act on sensible objects only through the intermediary of actuators on the body. The body is the point at which we sense and act on the world -- it is the mind's interface to the world. Even if the world is viewed as a set of images, the body is a special kind of image, a privileged image.

Thus, in our perceptual account there are not only sensible objects and a self that senses them. There is necessarily a third element -- the sensor -- that mediates between them. There is no sensation without a sensor. This fact immediately brings all the rules of sensor theory into play. The sensor output is formally and effectively determined by the stimulus or object, because the sensor is essentially passive and must be actuated by a stimulus. The sensor is a field of potential qualities which are selectively actualized by the stimulus. Put simply, sensations come from without.

We also discover that the world of sensation (unlike the world of imagination) is not entirely reducible to human sense qualities. Artificial sensors reveal to us new aspects of old objects, as well as entirely new kinds of objects. No one seriously doubts the reality of radio waves or x-rays, even though they are not directly observable by human beings.

There is also the perceptual fact of convergence: with increasing resolution, the apparent properties of a sensible object converge to a limit. Convergence occurs not only for complex qualities like shape and texture, but also for simple qualities. It is also seen in the spatio-temporal continuity of the world.

Now, what does this revised account of the world as "naively" perceived imply? It implies first that the sensory world and all the things it contains are *outside* us. They are not subjective products of the mind. This is due to the fact that sensation is impossible without the intermediary of a sensor and that the sensor must be actuated by an object. We do not will sensations. The body, with its sensors and actuators, is the boundary between "inner" and "outer". The regions given in sensory images are caused by things that are "out there".

Second, qualities are at least partially objective and correspond in some fashion to the properties of objects. Qualities, as sensor theory shows, are the "signatures" of objects as perceived by the knower. Qualities are actualized by a sensor in an information system only with reference to the object known.

Third, the world of sense is real. It exists independently of one's subjective perception. It is not reducible to human sense qualities, yet the fact of convergence shows us sensible things do have definite properties. It also shows us the world as an external spatio-temporal continuity.

To sum up the import of the preceding discussion: *the image is the interface*. It is the point of contact between the mind and the world. Qualities are the "forms of sensation", bearing in them the mark of the order of the physical world. In transduction, this order is translated into a symbolic, mental form: informational rather than material. The image, composed of regions of qualities and intensities, thus allows the overlap of the mental and the physical in a formal sense. Images are essentially bodily -- they are produced by sensors. Sensors and actuators are intermediaries between mind and world on the formal level. We can know and reshape the orders of things through them. As for the "matter of sensation" -- or what we would now call energy -- it can only remain unknown through sensory images, at least in a static sense. In that one respect, the image is an impenetrable screen of appearances. For the unaided human senses, with their limited field of qualities, the interface provided by the image to reality is limited. But artificial sensors can both expand and refine our image of the world and our depth of view into it.

9.5.2 SENSING VERSUS IMAGINING

The question of whether there is a real, knowable difference between sensing and imagining is closely related to the status of sense qualities. It is easy to see how the problem of sensing versus imagining arises, especially as it is normally presented. In our sensible experience, we perceive objects and regions in space that are characterized by qualities that come and go in time. But our imaginings are also characterized by spatial extent and qualities. So are illusory mental images or hallucinations. What, if anything, distinguishes the images we possess or produce in the mind from those we immediately sense? If one holds what we perceive are little more than impressions or ideas, then, indeed, there can be no difference of the two. This is why Descartes held there was no essential difference between the sensible experience of our waking lives and the images we experience in dreams. This question is also highly relevant today with regard to virtual realities -- at what point (if any) does virtual reality pass completely for reality itself?

For sensor theory, there are several knowable differences between sensing and imagining. (By imagining here is meant the review or invention of images within the information system itself.) It is knowable if new images are actually being produced by the sensors. This is certainly true for artificial sensors, but also for our own senses. We are aware of the fact that we are *sensing,* that our sense organs are in operation. As the previous section discussed, we are aware of sensible things as being outside us, and we do not will the stream of sensory images. Imagining, in contrast, requires a suspension of, or inattention to, the sensory stream. Sensor images also possess a direct continuity with what is sensed an instant later. For an artificial sensor, the frame rate can be varied at will, meaning that the sensory world has an inherent continuity independent of how the sensor chooses to chop up the stream of information. Also, we can act in the world of sensory experience, while we can not *act* in our imagination in the same way. We can, of course, daydream of doing as we like. However, if we attempt to mentally produce a detailed image and operate on it, we find we can do so only with great difficulty. Our imaginings are pallid and threadbare in comparison with sensations. The same is true of dreams. The seamless integration of sense perception and action typical of our exterior experience is lacking. (And there is no such thing as active sensing for imagination.) And, as we saw earlier, sensing is of the real, whose properties are not exhausted by human sense qualities, while the mental image as image is reducible to qualities alone.

Closely related to this problem is that of interior illusions or hallucinations. We saw in the previous chapter that exterior illusions do not negate the possibility of our knowledge of sensible objects. Now we must consider the interior illusion: a false appearance or image within the information system itself. In its full sense, this is a problem for psychology and cybernetics. Sensor theory does provide a foundation for the proper treatment of the problem, however. Interior illusions take several forms, but their distinguishing mark is that they are erroneous modifications or productions of sensory images by the information system, rather than an effect of the passage of the stimulus through medium or its reception by the sensor. Hallucinations are perhaps the best example: the proverbial alcoholic who sees pink elephants in a delirium. Several varieties of drugs (e.g., peyote or LSD) have a primary effect of inducing hallucinations and producing strange modifications (e.g., color changes) in ordinary sensation. Another perennial example is the phantom limb: people who have lost an arm or a leg have sensations as if the limb were still present. Occasionally, phantom limb sensations are quite painful and require special treatment. There is also the previously mentioned problem of synaesthesia, where sensations of one mode lead to perception of qualities in another mode: audial tones may produce the sensation of colors, for example.

Interior illusions, just like exterior illusions, have long been used as arguments against realist notions of sense perception. In fact, more strongly than exterior illusions, interior illusions seem to imply there is no difference between sensing and imagining, no distinction between the real and the imaginary. Common sense, however, must ask, if this is the case, then how did the distinction between illusion and reality, between sensation and imagination, ever arise in the first place? The retort is that "reality" is simply those images we ordinarily experience, and "illusions" are the extraordinary ones. But sensor theory shows us there is more to it than that, and that the common sense point of view is not false. How do we know illusions to be illusions? First of all, they invariably belong to a single sensory mode. The hallucinatory pink elephant is purely visual, and so on. This means that illusions lead to attributing contradictory properties between sensory modes, whereas in normal sensation, the senses concur. (We have already shown there are no general exterior illusions -- this also applies to interior illusions, because they lead to contradictions.) Second, interior illusions, just like exterior illusions, can be defeated by active sensing. And that is how they are usually known to be illusions. One cannot reach out and touch a visual illusion, nor can one act on them.

This said, there is little doubt that the human senses can be tricked by images -- we can mistake an image for a real object. A movie, for example, is a series of static frames projected sequentially at high speed; but it is perceived by us as a continuous motion. And if one watches a movie of the descent of a roller coaster car, one can actually have a sickening internal sensation of motion. But this simply proves, once again, little more than the limitations of the human sensory system, and does not touch on sensors in general. Consider a virtual reality, for example, with projected visual images, sounds, and so on. Let us even suppose it has smell and taste, and one is enabled to move objects around. Such a virtual reality might seem very real indeed to a human being. Yet even here it might fall short of the entirely real. The most difficult of the human senses to fool would be touch, to present completely authentic, tangible objects to be picked up, handled and felt. And that is precisely because *touch is the only active human sense*. (Think of how hard -- if not impossible -- it would be for a virtual reality to simulate food and make us think we are actually picking up, chewing, tasting, and swallowing "virtual food" as if it were the real thing.) This brings up an astounding point. The human senses could be fooled by a virtual reality; at least in theory, it would be possible to build a virtual reality that would be indistinguishable from reality. But it would never fool an active Argus sensor. Such as sensor could interact with the entire spectrum of physical interactions, meaning that to appear fully real to the active Argus sensor, *it would have to be real*[16].

9.6 INTENTIONALITY AND THE MUTUAL CONFORMATION OF SENSOR AND OBJECT

If qualities are objective and meaningful in the way sensor theory describes, it implies the relation of the object and knower is not simply one of cause-effect, but formal -- it is the one-way *intentional* relation. In other words, the validity of sensation is also tied to the problem of intentionality. Since intentionality is notoriously hard for many people to grasp, it is best to review briefly what it means.

Intentionality arose in the Aristotelian and Scholastic theories of cognition. The sensible species (i.e., data representations or qualities) are the form of the object, as conveyed to the knower, and provide the means of knowing particular things. The sensible species are the only way that the object can "get into" the knowing subject for Aristotle. Intentionality means that qualities exist in the knower only with reference to the object known. They are not just a bridge between knower and known, but a one-way bridge. The intentional relation is asymmetric of its very nature. It implies the conveyance of a form *to* the knower, and the corresponding awareness by the knower *of* the object, and not the reverse.

We do not perceive the intentional qualities or data representations as such. Rather, we perceive the object *through* them, just as in looking at the moon through a telescope, we see the moon and not the telescope. They are merely signs of the thing, a pure means for the mind to encompass the object known. They are about the object, not themselves. Intentionality implies an "extending out toward or including" (*intendere*) of the object by the knower[17]. For Aristotle and the Scholastics, perception involves a sort of participation of the knower in the form of the material being. The intentional relation alone allows the object to become part of the act of knowing. Conversely, without the intentionality of sensation, the bridge between the subject and the object is broken, and knowledge becomes impossible in Aristotelian philosophy.

We have seen that modern philosophy broke with this view of sensation. Qualities, especially the secondary (i.e., simple) qualities, became subjective sensations. In other words, we do not perceive things directly; what we perceive are subjective sensations, ideas, or impressions. These are related to the object not intentionally, but as effects to causes. This meant that sensations had no necessary connection to the nature of the object. And, although it was by no means the only factor at work in modern philosophy, this fact had astounding results. The abandonment of intentionality had the effect of cutting the knower off from the nature of the object, and thus led to the most vexing problems of how objective knowledge is then possible. Conversely, we have seen that some doctrine of intentionality is essential to a realist epistemology.

One of the most important developments of 20th Century philosophy has been the restoration of intentionality by Husserl and the phenomenological school. But if the older view of intentionality was from the side of the object, the phenomenological view is from the side of consciousness[18]. Intentionality is the hallmark of consciousness: human consciousness is always consciousness *of* something. It is an asymmetric relation, a vector. (We must note that, for phenomenology, the intentional act constitutes the object formally. The object belongs to consciousness really, not just intentionally. This makes it more similar to idealism than to realism.) What is important here is the reappearance of the intentional relation and qualities as meaningful "bridges" to the essences of objects.

Thus, if qualities are objectively valid -- correspond to the object in some way -- then they are intentional. What must be shown is that the conveyance of the form of the object, the contact (partial or complete) of the knower with the nature of the object known, is in fact a condition for the possibility of sensation. In other words, intentionality and sensation must necessarily go together. This can be shown from two different angles. First, assuming the validity and objectivity of sensation, we will examine what conditions make it possible. Second, we turn this around and ask what must the sensory process be like in order for qualities to be intentional. And we must also keep in mind the direct or symbolic nature of the relation for complex and simple qualities, respectively.

(1) Sensation is possible only when there is a stimulus, and when this stimulus meets the sensor's conditions: that is, when it conforms to the sensor's mode, band, field, and intensity range. Otherwise, the sensor can not receive the stimulus (as was demonstrated for the protosensor in Chapter 2). Similarly, a particular quality (data representation) will be output by a sensor only if the stimulus matches its particular frequency, spatial pattern, etc. The stimulus must conform to the quality's conditions in order to be sensed as that quality.

Thus, the sensory object (the bundle of physical interactions) must overlap with the sensor conditions. That is, in order for sensation to be possible, the sensor must be viewing the whole or part of a sensory object. And the real object must interact with the very same mode and band of physical carrier as the sensor. (We saw in Chapter 7 that a full spectrum of physical interactions completely expresses the physical properties of the object -- a property without physical interactions is not a physical property.)

This means that sensation is possible (a data representation can be actualized) *only* if there is a correspondence between the sensor's conditions and the properties of the object. In other words, there is a mutual conformity or complementarity of sensor and object when sensation occurs. And it also means that the data representation, sent by the sensor to the information system, has meaning only in reference to the properties of the object -- it points to these properties. Thus, it has intentionality: the data representation objectively stands (however imprecisely) for real properties of something in the real world outside the information system. These data representations, when perceived, are qualities. In the case of macrocomplex qualities, the representation is direct and "pictorial". For microcomplex and simple qualities, it is symbolic -- the perceived quality is a sign of something real. Thus, the objectivity of qualities -- however clear or vague it might be -- is an underlying condition of sensation. Some information about the object must be conveyed to the observer to actualize a data representation and make sensation possible -- the object must always leave its "fingerprints". A completely "empty" causing of sensation is impossible. (This fact is simply an extension of what was already proven for the validity of the protosensor.)

(2) Now we examine what the sensory process must be like in order to

possess intentionality. First of all, the distinction of stimulus and data representation (necessary for the sensor by definition) could not be drawn unless the physical stimulus was real and different from the data representation in its mode of existence. The intentional data representation is a *sign* of the object, not the object itself. This means that while the stimulus is physical or material, the data representation (as a symbol) is not *essentially* material: information does not depend on the material composition of its carrier. But for precisely this reason also, the object that is sensed must be in the world outside the information system and is communicated to it only through the intermediary of a sensor. Otherwise, the information system would have to physically incorporate the object to know it -- or remain ignorant of it altogether.

Second, the sensor must have at least some of the same physical interactions as the object. Only in this way could the sensor image possibly represent the physical properties of the object. To represent the object fully, the sensor would have to be receptive to the same extent, resolution, spectrum, etc., as the object itself. Third, the intentionality of the sensory object or the sensor image must be maintainable through an indefinite series of transformations and mappings. Chapter 8 showed that projections do not, in and of themselves, destroy our knowledge of the object. As long as the transformation or mapping is known, we can indeed trace the image back to the object.

Therefore, if the sensory process is intentional, it must be as sensor theory has posited in the previous chapters. The object must be different from the information system, and must act upon it through a sensor, which is essentially an intermediary between the physical and informational realms. And the sensory link must be akin to a communications channel, conveying information about the object to the sensor.

Active sensing also has something to add here. We have seen that for every sensory object, there is a corresponding active sensor that matches it exactly -- active sensor and object "resonate" with each other like two identical tuning forks. We have also seen that active sensing, of its very nature, establishes a closed circuit between the information system and the object. It is no longer a one-way action of the object on the sensor, but an interaction of the two. The command given to an actuator by an information system can be exactly the same as the data representation of a quality yielded by a sensor. Now, if the object "resonates" with the quality/command of the active sensor, there is obviously some correspondence of the quality and something in the object.

We can also examine this question briefly from the side of conscious perception. Conscious intention (attention to one part of the sensor image) is a narrowing of conditions. It means that the representation of the properties of the object *to* the subject makes consciousness *of* that object by the subject possible. Can we become aware of what is entirely unstructured and disordered? It seems not: consciousness of disorder is simply a contrast with order. This implies that form or structure is a precondition of awareness, at least in a static case. That would further imply that the qualities which sensory images are composed of must be ordered, or else we would be unconscious of them. Sensor theory says this order, at least in part, comes from the object. Thus, what grabs our attention can do so because of an objective order -- sensory attention will be dependent on the intention of qualities.

Intentionality brings us back to the mutual conformity of the sensor and the object. We touched on this back in Chapter 2 with regard to the validity of protosensors. We see it is also the foundation of the validity of all other sensors -- which should not surprise us, as all sensors can be represented as a combination of protosensors.

Mutual conformity means that sensation is rather like a key and a lock -- the object and its properties are the key; the sensor and its conditions are the lock. The Aristotelian theory regarded the senses as a formless wax, ready to receive impressions from all objects. In other words, the senses (and knowledge in general) conformed to the object -- the lock can be opened by any key. Kant, in his famous "Copernican Revolution, turned this around, and posited that the object must conform to the conditions of the subject. He was the first to seriously examine "the lock" and see that it does matter. But in doing so, he also rendered "the key" meaningless. Sensor theory means that conditions of both sensor and the object must match in order for sensation to be possible -- the key must fit the lock. A sensor, in essence, is a set or field of data representations that can be actualized by a stimulus when its conditions are met. Sensation occurs whenever a stimulus meeting the conditions impinges on the sensor -- that is, whenever a proper "key" is inserted into the "lock". A simple "lock" (e.g., touch) can be opened by any number of different "keys" (e.g., all solid objects). A more specialized "lock" (e.g., a hair cell of the inner ear) can be "opened" only by a few "keys" (e.g., a narrow range of sound frequencies). Intentionality and the possibility of sensation are joined together. Something of the form of the object, its "signature", must be conveyed to the knower in order to be sensed. We can see also in what way that sensation is informational rather than material. A lock doesn't care what the key is made of, only that it is of the right form. The sense organ detects a particular kind of stimulus, with a particular resolution (fine or coarse), and encodes that information in a fixed way. This encoded information -- the qualities of sense -- form the bridge from the knower to the object. They are the point of contact between the knowing subject and the nature of the known.

The sensor's *a priori* conditions define the "world" of possible objects it can encompass -- that is, the objects it can enter into a relation with. The object, in order to be an object for that sensor, must conform to its conditions. Returning to the information theory model of the sensory process, the sensor is like a radio that can be tuned only to certain stations -- those stations constitute its world. At the same time, the radio reliably conveys to us whatever is transmitted by the station it is tuned to.

CHAPTER TEN

IMAGE PROCESSING AND PATTERN RECOGNITION

10.1 PRELIMINARY

Image processing and pattern recognition form the horizon of sensor theory. Beyond them lie the regimes of artificial intelligence, the psychology of perception, and epistemology proper. The distinction between sensation and perception given in the first chapter is relevant here. Sensor theory is concerned only with sensation. Perception assumes a mind or conscious subject that does the perceiving. Hence, it involves considerations of the nature of mind (or the nature of the information system) not necessary to an understanding of sensors and image information.

Image processing and pattern recognition must be considered at least cursorily by sensor theory. Perception is based on sensation, but what we perceive does not usually consist of raw sensations -- that is, raw sensor outputs. Images are invariably "pre-processed" or "digested" before they reach the level of conscious perception, to enhance certain features of interest and suppress unwanted backgrounds. Some pattern recognition is usually involved, too. This is particularly true of human vision. A second point, related to the first, is that we perceive only a small fraction of what we sense, as has been shown by numerous studies. If we did perceive everything sensed, we would be overwhelmed by it. Thus, there is a selection of information from the image before it is perceived or as an integral part of the perception process. All these varieties of image processing do not belong to perception as such, as they are performed automatically and unconsciously. They are really the last stage of sensation.

What is true of human sensation also holds for artificial sensor systems. Artificial sensor systems typically pre-process images before they are analyzed for patterns. And sensors often also attempt to reduce, simplify, or compress the images, since the information system may not be able to handle the amount of incoming sensory information. The parsing of the data stream into discrete frames, as discussed in Chapter 5, is one example of this.

Certain aspects of image processing have been extensively studied. These include techniques of pattern recognition in general, and optical character recognition and speech recognition in particular. Students of human perception have been able to infer several information processing stages that take place between the sense organs and conscious perception. Radar and optical sensor systems generally involve locating and tracking moving objects. Artificial intelligence researchers have devised several schemes for the classification of images. Even so, the study of image processing and pattern recognition is still in its infancy, and it will doubtless grow tremendously in the decades ahead, driven by the demands of computer graphics, virtual realities, and text and speech software. In general, image processing has grown hand-in-hand with advances in computer technology.

In image processing and pattern recognition, there are several distinct points to consider. First, there are operations which take place on a single sensory image. Second, there are operations on multiple images. Next, there are the related questions of the analysis or segmentation of the image into distinct objects, attention to one part of an image, and recognition or identification of the objects found. Finally, there is the problem of the classification of objects, abstraction, and the formation of concepts. This last stage is the point where sensory information and thought come together.

10.2 OPERATIONS ON A SINGLE IMAGE

Operations on a single image are usually performed to enhance desired features or to favorably rearrange information for later analysis. Single image operations are a stage of preprocessing and may take place early in the sensory process (e.g., within the retina itself in vision). All single image operations have the form:

Output image = F (Input image)

where F is the image operation function. Single image operations may also be referred to as *filters*, the normal term used in signal processing. The number of possible filters is limitless. This section will concentrate on the ones most commonly used for image analysis. One fact concerning images should be remembered from the outset. Although we quite naturally perceive the world as analyzed into distinct objects, this is not the character of the image itself. The image consists of a spatio-temporal field of intensities and qualities. It must be considered *both* as whole and as parts. It is not a collection of sensory atoms. We tend to overlook this, because our sensory faculties preprocess images before we consciously perceive them. The importance of single image operations is just this: they make analysis of the image easier. In a way, they constitute a ranking of objects in image. Because certain features are enhanced, others must be suppressed. It also assumes some prior knowledge of what features or aspects are most desired.

One of the most important single image operations is *edge enhancement*. This is particularly true of vision. The primary purpose of vision is to detect the spatial properties of objects: their shapes, sizes, and motions. Solid objects tend to have sharp edges that appear discontinuous on the macroscopic level. And even liquids have sharp surface boundaries. Thus, the location of macroscopic objects depends integrally on the detection of edges: there seems little doubt that edges or contours are the key factor in the spatial organization of visual perception. Anything which highlights edges assists in the detection process.

A common means of edge enhancement is to take the spatial second derivative of the image. This is also called the Laplacean operator (∇^2), and, unlike the first derivative or gradient operator, it does not have a directional or vectorial character. The second derivative of a sharp edge has the effect of suppressing the object and leaving the edge only; the same operation on a smoother boundary will tend to create a demarcation. In both instances, the edge is marked by a zero-crossing in the second derivative of the image. It also has the effect of increasing the contrast between the brighter object and a darker background.

Human vision performs edge enhancement through an effective second derivative of the image, and it does so at a surprisingly early stage of the visual process[1]. The approximately 130 million rods and cones of the retina do not feed directly into the optic nerve and the brain. The retina has three distinct layers, and the layer of cells just behind the rods and cones set up horizontal connections, or fields, to which the third layer of ganglions responds. In this way, the outputs of the 130 million photoreceptors are reduced to approximately 1 million ganglion receptive fields. The optic nerve fibers are connected to the ganglions: hence, they each view of spatial field of receptors. The contributions of the photoreceptors in each receptive field are not simply added up, however. Rather, there are on-center fields and off-center fields. In an on-center field, the response to light in the center of the field is positive, while the response of the surrounding region is negative (Figure 10-1). The response of a function of radius is very close to being the second derivative of a Gaussian, which is precisely the operator used to numerically take the second derivative of images. In other words, the second and third layers of the retina are taking the effective second derivative of the image for purposes of edge and contrast enhancement. Edge enhancement is not without its price, however. It is the cause of the familiar optical illusions shown in Figure 10-2.

Another single image process, related in many ways to contrast enhancement, is *thresholding*. Thresholding means that everything below a specified threshold intensity level is set to zero and ignored. This completely suppresses the background below the threshold level, leaving only bright regions or objects of interest. Thresholding will be of value when the objects of interest are significantly brighter than a structured background. Thresholding then allows the information system to focus on the object and its properties alone. A series of different thresholds will yield the *intensity contours* of the image.

(a) visual receptive field organization in the retina

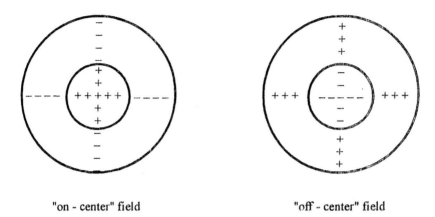

(b) second-derivative Gaussian operator (on - center) for edge enhancement

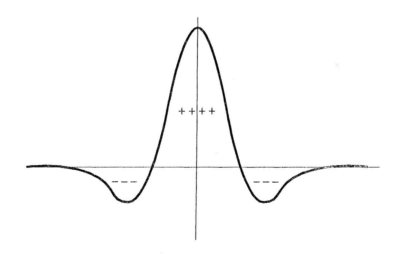

Figure 10-1. Visual Receptive Fields and the $\nabla^2 G$ Operator.

The Hermann Grid

(gray dots at intersections are an illusion from edge enhancement)

Mach Bands

although the circle is a uniform gray, it appears to have a lighter band at its edge.

Figure 10-2. Optical Illusions Resulting from Edge Enhancement.

Thresholding and edge detection are both indispensable for the figure-ground organization of images[2]. In most visual images, we perceive certain parts of the visual field as a sharply delineated, cohering object or region, while the rest of the image is perceived to belong to a less important background. The figure perceptually stands out against the ground and draws our attention. This is almost entirely due to the pre-processing that occurs in the visual system long before the image is consciously perceived. Our vision, in other words, is hardwired to locate bright, well-defined objects and draw our attention to them.

Thresholding may also be considered a special case of the more general class of image operations known as *normalizations*. An image may contain variations which are unnecessary for the identification and classification of patterns, and may actually interfere with such identification. Rather than developing a classification scheme that attempts to deal with these variations, it is advantageous to remove them prior to classification[3]. Normalization is thus a method of simplifying images. One common example is the smoothing of images. In smoothing, intensity values from several adjacent pixels are averaged. The image may in fact be rebinned from higher to lower spatial or temporal resolution. This removes small-scale detail and structure which may be superfluous in the identification of larger shapes and patterns. There are many such averaging schemes; a detailed discussion of them belongs to a text on image processing. Here it is enough to note that normalization means the removal of extraneous information: unwanted background or noise according to the purposes of the information system or organism.

Perhaps the most powerful of all single image operations is the Fourier transform. This was touched on in Chapter 3. The Fourier transform has the effect of transforming the frequency structure (the resolution or intension) of an image into extension, and vice-versa. Figure 10-3 contains some examples of transformed images. In time, this means we obtain a new image that contains the frequency composition of the old image. The Fourier transform of a single musical note would be a series of spikes, largest at the primary frequency and smaller at the overtones. The transform of a chord contains a larger series of spikes, while the transform of human speech is quite complex. For spatial images, the Fourier transform yields a new image of spatial frequencies. The transform of a grid, for example, is a rectangular field of dots. In general, the

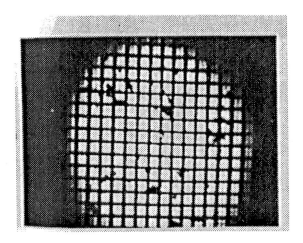

Original Image:
A dusty mesh illuminated by a laser

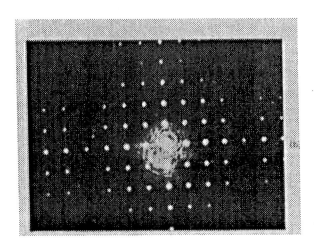

Transformed Image:
The transformed grid results in the regular pattern of dots, while the dust causes the irregular pattern in the center. It is now easy to separate the image of the grid from that of the dust

Figure 10-3. Examples of Images and their Fourier Transforms.

entire effect of a Fourier transform is to rearrange the information in the image by pattern. This is obviously of great benefit in pattern matching and pattern excision.

Some effective Fourier transforms take place in human sensation, in that some visual neurons are apparently tuned to different spatial frequencies, and hearing is very sensitive to the frequency structure of sound stimuli. Use of the Fourier transform in artificial sensors has been made possible mainly by advances in computers and in Fast Fourier Transform (FFT) algorithms. It will doubtless find steadily wider application in image processing in the future.

10.3 IMAGE SEGMENTATION AND ANALYSIS

It is rare that an information system will use an entire image, as a unbroken whole, without analyzing it into figure and ground or locating objects within it. (Appreciation of beauty is a rare example, as it involves a holistic view of an image.) More usually, one of the main purposes of the information system is to isolate objects and patterns of interest contained in the larger image. In the visual field, we naturally seek out distinct objects; we just as naturally divide what we hear into distinct words. This everyday process is image segmentation or analysis, the breaking up of an image into a number of sub-images. Most of the preprocessing techniques of the previous section are meant to assist in just that. Also, the separation or location of objects must precede the identification of the objects. The reasons for this are obvious: we must determine if something is there before we determine precisely what the something is.

Segmentation or analysis of the image is necessary not only for locating objects. It is also needed due to the limited information processing capacity of both computers and human perception. We perceive only a small fraction of what we sense, as Table 10-1 makes clear. Image analysis filters from the great mass of sensory information that limited amount that is most important (e.g., the approximate centers of objects). And usually the desired information *is* much less than the contents of the entire image. (These limitations would not pertain to a computer with unlimited storage and unlimited time to analyze the images.) The reduction of information in the image to what is truly necessary is due to our need to act and respond to that information in a timely way. Put another way, the faster the response must be, then the more simplified the information needed for the response must be and the more automated the information processing system. Conversely, reflection and deliberation requires time and a release from the necessity to act.

Table 10-1. Comparison of Sensory Information with Perceived Information[4].

sensory mode	maximum information of sensory image	maximum information perceived
vision	10^7	40
hearing	10^5	30
touch	10^6	5
taste	10^3	1
smell	10^5	1

Two possible levels of analysis can be identified in sense perception. First, there is the entirely unconscious preprocessing of images before perception. Figure-ground separation is an example. Such processing is really a sort of computational algorithm for pattern location. Second, there is the conscious attention to an object or part of an image. This selectivity is part of conscious perception itself.

Let us consider first the pre-conscious segmentation of images. Techniques for segmentation of images into objects and patterns have been developed for artificial intelligence applications; there has also been much progress in the understanding of what goes on in human visual processing. Segmentation of images is not something simple; it may require several processing steps. Complicating matters in the case of human perception is that we cannot directly observe all aspects of the sensory process.

The most fundamental assumption of all image segmentation is that objects (regions of interest) will be characterized by commonalities, while the boundaries between regions will be characterized by differences. Differences must exist within the image for analysis to be possible. There must be separable patterns, regions, or groupings of objects. Otherwise, the entire image is just one big pattern or object. Image analysis thus assumes that the world is "lumpy" and articulated into discrete objects. The more vague the transitions between patterns in the image -- the more hazy and cloudlike they are -- the more difficult and less meaningful separation of distinct objects becomes.

In hearing, phonemes and syllables are identified as units and separated out on the basis of pitch, loudness, and duration of the pattern. Commonalities of visual regions include texture, shading, shape, common motion, and color[5]. In other words, regions are characterized by similarity of qualities. Of these, spatial frequency structure (texture, orientation of surface patterns, etc.) seems to be the most important[6]. Such structures or patterns are microcomplex qualities -- we do not normally consciously perceive the patterns and their spatial scales as such, but they are cues to the spatial organization of objects within the images on whose surfaces they appear. Indeed, the real utility of microcomplex qualities is in image analysis.

These facts square nicely with the perceptual rules of Gestalt psychology[7]. Gestalt psychology holds that we do not perceive objects as additive composites of sense-atoms, but in a holistic fashion: objects perceptually cohere *as wholes*. There are several factors that govern the degree of coherence of patterns and regions, such as proximity, similarity of shape, symmetry, good continuation (smoothness), common movement, and closure. In other words, coherent patterns have a common spatial organization and relative simplicity. It is thought by some that human perception segments visual images into a relatively small number of kinds of volumetric objects (blocks, cylinders, wedges, cones, etc.) on the basis of symmetry, parallelism, collinearity, curvature, and cotermination of lines[8].

David Marr, in his book *Vision*, outlined several analytic steps that occur in human visual understanding[9]. Starting from the source image, the visual system first composes what he calls a primal sketch. The purpose of the primal sketch is to determine the changes and structures of the image. Apparently, the primal sketch contains the results of second-derivative edge detection (discussed above in section 10.1). It also contains "place tokens" that correspond to oriented edges, bars, line terminations, and blobs. These contain the essential information needed to interpret the geometric structure of the objects in the image. (There is good physiological evidence that different spatial features of a pattern excite different layers of the visual cortex in the brain -- i.e., a visual pattern is indeed analyzed into oriented edges, line termini, etc.[10]) The next stage of visual processing yields what Marr calls a 2 1/2 dimensional sketch. This is a representation of the geometry of the surfaces, as based on such factors as local surface orientation, distance from the viewer, discontinuities in depth, and discontinuities in surface orientation. In the 2 1/2 dimensional sketch, depth information is extracted, in other words, from the geometric information contained in the primal sketch. Only after these two preliminary processing stages are visual objects or patterns recognized, and the geometric aspect of this 3-D recognition takes place in a schematized way. The point is that analysis of the visual image is a complex process; much takes place before we consciously perceive something. And this process is one of simplification, of extraction from the image of that information that is truly necessary. There have been attempts to apply Marr's visual paradigm to robotics and computer vision[11].

Segmentation of images cannot be, however, entirely separated from pattern matching[12]. The reason is simple: images can be very complex, and an uninformed segmentation of patterns correspondingly laborious. If one knows what patterns to look for, segmentation becomes much easier. Context is also important. It may be that segmentation and pattern recognition is necessarily an iterative process. The exact combination of processing steps depends on the purpose of the system employing the sensors.

This brings us to the second kind of image analysis: conscious attention. Attention is the selective property of perception. It means "looking at" or "listening to" one part only of the full sensory image available. (This is necessary once more because we perceive only a fraction of what we sense.) Attention means focusing on one object or one action. Thus, it is really a reduction of the image rather than its segmentation. But, in order to find a object, it must rely on segmentation criteria. In attention, the separation of objects from their background is part of perception itself.

Attention, of its nature, divides the image into focal (liminal) and peripheral (subliminal) zones[13]. The two dimensions of attention most relevant to image processing are *concentration* and *search*. Concentration is the exclusion of sensory stimuli that interfere with awareness of the desired pattern or object. It means, in other words, the suppression of the peripheral zone as much as possible. Search involves looking for a specified stimulus among many stimuli. It requires awareness of the entire image. It also requires pattern recognition -- attention is directed towards the desired pattern. In general, conscious attention needs pattern recognition as an integral part of the selective process; it is not a mechanical and uninformed segmentation of the image. This appears to be true even when we are not specifically looking for something, and a stimulus spontaneously "catches our attention". Experiments have shown that size, intensity, and motion are important determinants of "eye-catching" images[14]. Another attracter of spontaneous attention are novel, surprising, incongruous, and moderately complex collections of objects or patterns -- a fact well-known to advertisers.

What, then, is the cognitive result of image analysis? It is this: one or more "somethings" are "out there", with extents and locations in space, durations in time, and various sense-qualities. In other words, analysis tells us *that* something is present before we know exactly *what* the something is. At the same time, we see that pattern recognition cannot be rigidly separated from image segmentation. If we know what we are looking for, it becomes much easier to analyze an image. The danger is that we might impose the desired pattern on an image in which it is not present, and thus perform a faulty image analysis.

10.4 MULTI-IMAGE OPERATIONS -- THE COMPARISON AND CONTRAST OF IMAGES

Before exploring pattern recognition as such, we must first consider operations on multiple images, because pattern recognition depends upon the comparison of an image to some reference pattern, either directly or indirectly. Multi-image operations all amount to the comparison and contrast of images on the basis of their internal similarities. The output of a multi-image operation can be a scalar number (e.g., the measure of the similarity or correlation of the images) or it can be a new image:

(1) $A = O(I_1, I_2, ..., I_n)$ scalar measure
(2) $I' = O'(I_1, I_2, ..., I_n)$ new image

Some multiple image operations will require a specific number of input images (e.g., the time differencing of two successive images) or can accept an indefinite number of inputs (e.g., addition and averaging of images).

The comparison and contrast of images assumes that some basis exists by which to meaningfully associate them. Classical associationist psychology determined three basic rules for the association of images. First is association on the basis of contiguity -- similarity of location in space. Second is succession or cause-effect -- the closeness of two images in time. Finally is resemblance -- similarity of qualities. Sensor theory provides a basis for all these three kinds of association. They are simply three possible different kinds of image similarity and difference. These similarities -- spatial, temporal, and qualitative -- are the same as the grounds for the synthesis of complex sensory information (Chapter 3).

Association of images depends on some measure of similarity -- a measure of the "nearness" or "farness" of images to each other in space, time, and quality. There are several possible approaches to measuring image similarity. The difference between two different pixels can be calculated as:

I. spatial difference $\quad \nabla r = ((x_1 - x_2)^2 + (y_1 - y_2)^2 + (z_1 - z_2)^2)^2$
II. temporal difference $\quad \nabla t = t_2 - t_1$
III. qualitative difference
    A. intensity $\quad \nabla I = |I_2 - I_1|$
    B. qualities $\quad \nabla q = \Sigma |q_{i2} - q_{i1}|$

The problem arises in how to combine these different measures into a single "distance" between two images. It all depends on which is more important, and that may depend on the application. Each measure in itself is independent of the others[15]. The difference could be represented as a vector $[\nabla r, \nabla t, \nabla I, \nabla q]$ rather than a scalar distance. A more serious problem arises when we consider that the difference/similarity of interest may be between different objects in an image, or between the same object at different times or positions. Motion of an object is an example. Another is the comparison of the shapes of two objects. This would seem to presuppose segmentation of the image, and perhaps even recognition of the object, before the comparison is made. Another, perhaps more sophisticated, approach to image differences is cross-correlation, which involves the convolution of two images. However, cross-correlation still must deal with the independence of the spatial, temporal, and qualitative "dimensions".

All sensory images have some spatial aspect, some temporal aspect, and a set of mode-dependent qualities. Thus, all sensory images have some degree of "associability" with others, however distant or vague. Even for different modes, the spatio-temporal aspect remains as a basis for association, as we saw in the problem of multimode sensor fusion. The question now is if this inherent associability of images can be objective or not, and whether sensor theory alone can answer that question. One thing to keep in mind is that when images are associated, they form a quasi-symplectic space -- which is a way of saying that each image potentially could be perceived as equally near to every other image.

Association of images in space and time will be objective, as long as images from the same sensor are being associated. In the case of multiple sensors, there can be objectivity if their locations are precisely known. If all the images are in the same coordinate system, then the spatial differences of pixels and objects can be calculated. They are associated because they come from the same spatial field. In the case of time, a series of frames from a given sensor are associated because they were previously extracted (sampled) from the same continuous data stream -- before the association, there was an analysis. The association is objectively in the series of frames, and no inference of cause-effect is required. The comparison of spatio-temporal patterns (e.g., shape, envelope) will rest on a similar basis.

Qualitative association is more tricky, as qualities are often held to be purely subjective. We have previously discussed this point. In resemblances of simple qualities, it is clear that only like qualities can be compared: colors with colors, scents with scents, etc. This is also the key to their associability: colors can indeed be objectively compared as colors. The perceived likeness/difference of two colors may indeed be subjective. But the fact that we can compare and contrast them at all is underlying and objective.

Differencing is not the only multi-image operation. Another one is averaging. If we average an image with itself, we get the original image back unchanged. If we average two images, we get an image "halfway" between the two. Now, consider the possibility of averaging a large number of somewhat similar images (human faces, for example). What will emerge from this operation is a generalized, "blurry" image -- a template or model -- that forms the "center of gravity" of all the images involved. The individual features will be smeared out, leaving only the commonalities. Image averaging contains the rudiments of abstraction and generalization, something we will return to below.

We now come to a rather interesting fact. The totality of one's images, by their inherent affinities and relations, form a universe of images: a *world image*. This is true whether we consider images in human memory or a library of images stored on a computer. This points to a perfection of sensory knowledge: the continual expansion of the world image in breadth and depth. Sensory knowledge -- based upon the matching of images -- is approximate and something entirely different from conceptual knowledge. As we sense and act and live -- partake in everything that constitutes sensory experience -- we continually add to our world images. The exact contents of the world image will vary from subject to subject (information system to information system), because it depends on the particulars of the sensory experience of that subject. If one's sensory world image is narrow but detailed, we call it craftsmanship, connoisseurship, or technical expertise. A broad, generalized grasp of the world, on the other hand, is generally called "common sense".

10.5 PATTERN RECOGNITION AND OBJECT IDENTIFICATION

All pattern recognition or object identification involves two closely related aspects. Pattern recognition, first of all, requires the comparison of the pattern or object of interest to a *reference pattern*. This comparison is something completely different from the origination of reference patterns. (This latter belongs to abstraction, discussed below in section 10.6.) The recognition process can be in stages, a sorting process that sets up an inference hierarchy. It is usual, for example, to segment the image into objects (identifying them as "things which are present") and locate them spatio-temporally before determining what kind of thing they are. Pattern recognition next requires the corresponding classification of the pattern or object under the reference pattern class name. A "class" for now will be thought of as a collection or field of possible patterns.

As emphasized before, pattern recognition is a vast and growing field, driven by speech recognition, optical character recognition, and so on. It will find applications in the years ahead in video processing and information filters on computer networks. Given this fact, all we can do here is summarize a few salient points. Further study in this area should be directed towards appropriate specialized texts.

First, what is the purpose of pattern recognition? We may say that all knowledge is in some way based on it. But the primordial role of pattern recognition is to guide or trigger action. In simple animals (and in artificial robots), sensory images lead to recognition, and recognition leads to a specific action. A given pattern evokes a given response. The response may, in fact, be "hardwired" and reflexive. The shorter and simpler the pattern recognition process, the quicker the possible reaction. In some animals, pattern recognition must occur very early in the sensory process to ensure a rapid, reflex-like response to the stimulus pattern. Marr relates the example of a spider that has a V-shaped retina, which matches the red V pattern on the backs of potential mates[16]. Frog retinas are capable of recognizing small moving spots and signaling this fact to the frog's brain; the utility of this in catching flying insects is obvious[17]. In such cases, sensation, recognition, and response are telescoped into a single event, rather like flipping a switch and having a light turn on.

Human intelligence and human-operated systems are clearly not this simple or deterministic. Human beings are not robots programmed to respond in exactly the same way to patterns. For this reason, pattern recognition must be treated as distinct sensory phase apart from any response. That is why pattern recognition yields a classification of the pattern (P is R) rather than an active response (if P, do X). The purpose of pattern recognition is thus classification. Any pattern recognition scheme has as its aim to take a stimulus pattern and return its class name if it is "close enough" to the reference pattern of that class. For a class, such a recognition function is also called a perceptron[18].

There are three basic strategies in pattern recognition[19]. The first is to check each object (previously segmented from the entire image) against the reference patterns or models. This is known as a bottom-up or data-driven approach. The second is to take each model and try to find patterns that match it in the (unsegmented) image. This is the top-down or model-driven approach. Finally, there is the hypothesize and test approach, which combines the first two. Which of the three basic approaches is most advantageous will depend on the problem.

Niemann distinguishes between what he calls simple and complex patterns[20]. A simple pattern is one that is matched to yield a class name. A complex pattern, on the other hand, contains a number of patterns and must be segmented into component simple patterns which are identified.

In human pattern recognition, there are two basic hypotheses. (They are hypotheses because pattern recognition in the mind is not directly observable.) The first is known as the distinctive features theory[21]. It is theorized that the stimulus pattern is analyzed first into features, and it is the pattern of most prominent features which is matched. The perceived complexity of a visual object depends on the number of sides or angles it has. The second is the template-matching theory. Patterns are stored in memory as generalized prototypes or templates, which are then compared to the entire object. It seems fairly certain that we recognize three-dimensional forms in terms of remarkably simple and schematized models[22]. The fact that we readily associate even very simple line drawings with the objects they represent is evidence of this. At the same time, human recognition of objects depends to a good degree on context. We expect to find an object of a given kind in certain relations to others: a chair at a table, rather than entangled in the upper branches of a tree. If the normal context is altered, the ability to recognize objects is severely impaired[23]. We locate and identify smaller patterns with some assistance from the "big picture" as experienced in the past. Under constant conditions, that is doubtless one of the most efficient methods of recognition.

Pattern matching may be precise or vague. There will in any case be a "distance" between the pattern of interest and the reference pattern, as discussed in the previous section. The notion of a pattern belonging approximately to a given class has been extensively investigated over the past three decades by the theory of *fuzzy logic*[24]. In fuzzy logic, the result of a logical operation is no longer a binary true/false answer, but, rather an approximation that runs from 0 (false) to 1 (true). Fuzzy logic was invented to attempt to deal with the vague, approximate, or statistical nature of much that we deal with in the "real world".

If the notion of fuzziness is applied to pattern matching, one no longer has a hard and fast yes/no for membership in a class (i.e., match to a reference pattern). Rather, there is a degree of membership: a membership value. In the most general form, one can define a pattern matching function (class membership function) m(R,P) as:

$$m(R, P) = \begin{cases} 1, & \text{if match of R and P is exact} \\ \text{between 0 and 1, exclusive, if the match is approximate} \\ 0, & \text{if there is no match} \end{cases}$$

where   R is the reference pattern and
P is the stimulus pattern to be classified

In other words, m is a measure of the similarity of the two patterns, the probability that P is R. We see that pattern recognition is thus not always exact. It depends on a judgment of how close the value of m must be to unity before we classify P under R. (If R is a "blurred image" or template, the m value will be the standard deviation of the images that went into making R.)

10.6 ABSTRACTION AND CONCEPT FORMATION

The questions of the nature of abstraction and concept formation is one of the most controverted in the history of thought. It does not belong to sensor theory as such. But what sensor theory has to say regarding the relation of images to objects, and the foregoing discussion of pattern recognition, does have important implications for abstraction and concepts.

Abstraction has often been regarded as something mysterious or incomprehensible. It is not my aim here to review the various theories of it. All would agree, however, that abstraction means the separation of one or more common features or aspects of different things. It is the collection, unification, or synthesis of such common features. This set of attributes is then signified by a concept and verbally represented by a word. For example, when I see a house, and recognize it as such, that is because it fits certain attributes that are subsumed under the concept "house". The particular house belongs to the class "house". (The relation of the particular object and the general class is the great point of conflict between nominalist and realist notions of concepts.)

We see immediately that abstraction is something different from mere pattern recognition. Abstraction means the formation or synthesis of a new reference pattern, not the matching of an existing reference pattern. It means the definition of a new concept, not classification under an old one. Attention has its role here, too. In abstraction, we pay attention to certain features of objects and suppress others. Abstraction is thus one part of conscious perception and in this respect goes beyond what sensor theory can describe. Because what causes one to pay attention to certain features and not to others -- whether this be arbitrary or objective -- cannot be understood on the basis of the sensible features alone. There are different ways of seeing the same object.

It is clear then that concepts define classes, and they stand for properties or qualities held in common by several objects. Or, they may stand for the properties or qualities themselves. Abstraction means the segmentation of images in terms of type or class, rather than in terms of spatio-temporal objects. Here is worth returning to the distinction of extension versus intension. The extension of a concept is the number of possible objects (the area of "pattern space", if you will) it can cover. This is directly analogous to the spatial extension of a image. The intension of a concept is the number of attributes which define it -- its degree of detail. This is analogous to the intension (resolution) of the image. If we transform intension into extension (and vice-versa), we obtain an image reorganized on the basis of pattern, quality, and information. A segmentation of this image would yield "transform objects" that are sets of attributes. It may be that abstraction and concept formation takes place in this way in the mind.

(Part of the confusion here is doubtless due to the fact that while we perceive objects primarily *visually*, we think about them in terms of *auditory* symbols: words. Thus, in thought we are continually forced to go mentally back and forth between different sensory modes. One advantage mathematics has over all other fields of thought is that it works with purely visual symbols.)

We saw in section 10.3 that sensory images have an inherent associability, based on time, space, and quality. Since the sensible world is "lumpy" -- articulated into discrete objects of recognizably different types -- the "world image" will be organized into distinct image clusters. A model or template can

(a) objects in an image: a collection of regular polygons

(b) pattern clusters in the image

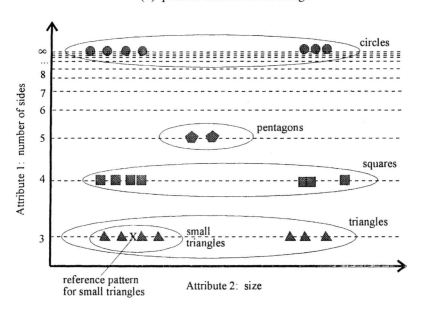

Figure 10-4. Cluster Analysis and the Formation of Reference Patterns.

be obtained for each cluster by averaging. Such an averaged "blurred image" represents the commonality of the cluster, its center. Abstraction would thus take place by isolating each group of related patterns and then averaging the patterns in each group to yield the reference pattern. Figure 10-4 graphically describes this process. It may also be that concepts and reference patterns are fixed in our minds by a single representative defining experience, where one image is taken as representative of an entire class[25]: "We learn the word 'pig', as we learn the vast majority of words for ordinary things, ostensively -- by being told, in the presence of the animal, 'That is a pig'."

Concepts signify each reference pattern and the attributes common to the images. We see that concepts will thus have a synthetic origin: they depend on a synthesis of attributes belonging to different images. We also see, however, that concepts obtained in this fashion can only be approximations: they are an average of images empirically observed to be similar to one another. If abstraction is merely the location of clusters in the world image, and concepts stand for the common features of the images found in each cluster, then one has a nice empirical-nominalist theory of abstraction and concepts. Concepts are experientially-formed approximations with an irreducible "fuzziness"; abstraction is an information-processing algorithm. Concepts are just the result of the segmentation of the world image by common features. The reference patterns to which the concepts refer would be the pinnacle of sensory knowledge. Doubtless, some process like this can account for the concepts of objects we sense and act on in our everyday lives, and how we recognize objects in experience.

There are a couple of serious problems with this view, however, if it attempts to explain all concepts. It cannot account for the idealized or "pure" objects on which mathematics is based, for example. In no sensory image do we ever find perfectly straight lines and round circles. We merely see approximations to these: a circle traced in chalk on a blackboard, for example. Yet mathematics must be absolute and ideal, or it is nothing. Which is the same thing as saying that the approximate, "more or less" nature of empirical concepts cannot yield necessary and universal truth. Yet we would all agree that mathematics is valid. This means there are concepts which are not approximations, but exact. Where such concepts come from is for the philosophic schools to argue; there is no doubt, however, that we possess them and that they cannot come *merely* from image clusters. In human thought and perception, there is the capacity to see the perfect and ideal in the imperfect image. We see the perfectly straight, infinitesimally thin line in the drawn pencil line on paper. Moreover, we can mentally extrapolate that line to infinity. This capacity to discern the perfect in the imperfect, to extrapolate to the infinite, and to interpolate to the infinitesimal cannot be explained on the basis of the images themselves. It is another kind of abstraction altogether.

We have also seen that the combination of images presupposes their underlying associability. Corresponding to this are concepts which are implied by any and all images, because they stand for the properties common to all images. These concepts can be grasped from a single experience -- no synthesis of images is necessary for them: to know a color, I need only see it once. These concepts are just those familiar ones that have emerged from sensor theory itself, the grounds of association and synthesis discussed above and in Chapter 4. First of all, are space and time, which are present in all sensor images in one way or another. Second are the modal intensities, such as brightness and loudness. Third are the modal quality genera, such as color and pitch. Finally, there are the names of the sensory modes themselves: vision, hearing, etc. None of these can be gotten from a combination of images, because they are the preconditions of such combinations. They are, so to speak, the different dimensions or axes of the world image.

Sensory information is therefore not all there is to knowledge and thought: mind cannot be entirely reduced to images and image operations. The image is, rather, the point where mind and matter come together in conscious perception. It is thus also the point where sensor theory is completed.

CHAPTER ELEVEN

SENSIBILITY IN GENERAL

11.1 THE PROBLEM OF SENSIBILITY IN GENERAL

Finally, almost as an addendum, this chapter considers sensibility in a completely general way. By completely general is meant an examination of the conditions of the sensory process for *any* possible universe or system, rather than just our own, with its particular set of possible sensory modes, limitations, and conditions.

Lest this be seen as an idle exercise, of interest only to writers of science fiction and the most abstruse of metaphysicians, there are actually some excellent *practical* reasons for exploring sensibility in general. A sensor was defined as a device that converts energy into information. But "energy" in the most general sense need not be limited to our particular universe, space-time, or physics. The core aspect of sensation -- the extraction or transposition of information from the stimulus -- can occur in several "real world" situations that are not spatio-temporal in the ordinary physical sense.

A good example is a station on a computer network which monitors activity on that network. The network has a multiplicity of interconnected nodes, but it is not a three-dimensional space. Our station can "see" a number of other stations, but this "vision" is not of a continuous two-dimensional field of view, but of discrete nodes.

Yet this can be considered sensation, because a carrier is translated into an internal data representation of the usage.

Multidimensional virtual realities are another possible application. It is entirely possible to present a four-, five- or N-dimensional world to a human participant in a computer-generated virtual reality. This might be of utility of understanding complex systems. The human observer would, of course, still have the same sense organs with the same properties. But the structure of the world which the subject observes and acts in will have completely changed.

Another example are organs in the body. An organ will typically receive signals from the brain, but it will also sense chemical substances in the blood. The interrelation of the organs through the nervous and circulatory systems once again is not a three-dimensional geometry (although it is contained within a three-dimensional physical space, of course). Yet the orderly reaction by an organ to chemical substances implies a kind of sensation. It is akin to smell or taste, although it never rises to the level of our conscious attention. Another possible example are economic systems, where business enterprises are trying to collect as much information as possible about market conditions and potential consumers of their products.

We see something very interesting emerging here. Sensor theory, considered generally, begins to merge into systems theory. A sensor is what allows a given part of a system to know what is going on in the rest of the system. (An actuator is what allows one part of a system to act on the rest.) Our part of the system is related to, connected to, in communication with, other parts of the system along various paths. The question becomes, then, how do we distinguish sensibility in general from other possible kinds of relations of systemic elements?

11.2 SENSIBILITY AND SYSTEMIC RELATIONS

Whenever one raises an important intellectual question, someone else has usually gotten there first. And so it is with the question of sensibility and relations. Kant drew a distinction between sensibility and intuition. By sensibility, he meant the possibility of sensing something: the possibility of being in relation to objects through sensation. But Kant did not use the term "intuition" in the way we do colloquially or even the way the philosophers normally do. By it, he meant the relation of the knower to the object in general. In other words, sensibility was just one kind of intuition -- a relation to particulars. (Kant also supposed there could be "intellectual intuition", a direct intellectual grasp of objects, although he denied human minds have this power.)

Sensibility, as understood by general sensor theory, is a relation to *exterior* objects. (I.e., sensor theory discounts the notion of an "inner sense" -- imagination is not sensation.) Sensations come from without. Whenever there is the relation of knower to known through the translation of a sensor (which is located on the boundary between "inner" and "outer"), such a relation may reasonably be termed sensory. Several possible examples of alternate kinds of sensation were given in the previous section.

There is a connection, as we have seen, between the theory of sensors and the theory of systems. Sensors and actuators are interface points between an element of the system and the system as a whole. They are one condition of the possibility of communication with other elements. They translate "signals" or "stimuli" from the surrounding "world" or "network" or "bus" into symbolic representations used by the system element internally. It is this translation that is the difference between "inner" and "outer". We can readily infer, then, the other requirements for sensation from the generalized structure of systems, and distinguish sensation from other kinds of communication.

First of all, sensibility requires the possibility of a multiplicity of objects to be sensed, even if they are of exactly the same kind. If there are no objects, then, trivially, there is nothing to sense and no possibility of sensation. If there is one and only one object to be sensed, its information can be fed directly into the information system without the interface of a sensor. The translation of energy into information typical of sensors thus implies the possibility of sensing a multiplicity of objects. This is true even in the case of a sensor designed specifically to sense one object -- the thermometer in a car's engine, for example. The thermometer, although connected to the engine of one particular car, could perform exactly the same function in another car of the same kind. As a thermometer, it would probably reliably translate temperature into an electrical signal even if removed from the car altogether. It therefore could *potentially* sense a multiplicity of objects, although it may never actually do so. Sensitivity to even one kind of stimulus means sensitivity to an indefinite number of possible objects.

Second, there must be a possible connection between the objects to be sensed and the sensor by at least one kind of carrier of information. The sensor and the object must share an interaction; the stimulus must be able to act upon and make a change in the sensor. In other words, there must be at least one "mapping" between the object and the sensor. This step of translation is, to repeat, essential to sensation. At the same time, the connection itself must be present to actualize sensation. The connection or relation of the sensor and the object may assume a continuous spectrum of possible magnitudes, meaning they are related in space. Or, there may be only a few discrete possible locations for sensors and objects, as is the case on a network. In any case, there must be a relation of object and sensor through one or more carriers for the relation to be sensory. Sensory relations are always spatial in some way.

Sensation, as was shown in previous chapters, means a kind of isomorphism or complementarity between the object and the sensor. They could not interact otherwise, and sensation would not be possible. But both sensation and action, requiring interaction through an intermediary, imply a secondary kind of isomorphism. Complete isomorphism or mutual conformity of knower and known would be direct and unmediated. It would indeed be a kind of "intellectual intuition". Thus, we see that "intuition" -- which is to say knowledge of any sort -- requires an isomorphism. The knower participates in the known in some way. *The special mark of sensation is that it forces the relation of knower and known into specific channels requiring a translation.* It is these channels that determine the possible interactions of the knower and known, and, in reality, define the formal structure of the "world" which the knower and known share. Chapter 8 discussed how we can understand sensation in analogy to a communications channel.

The final requirement for sensibility, and the most subtle, is the "need to know". What is the purpose of a sensor? It is to gather information about the surrounding world. The need to know is thus the need to know about the surrounding world. For external sensors, this is literally true; for internal senses, it means the body. This pragmatic information-gathering purpose of sensors is fairly obvious: if we had no reason to be interested in the world "out there", then why have sensors or sense organs?

Sensation, as Aristotle pointed out long ago, is necessary for activity in the world. Animals thus all have sense organs of some sort, while plants are insensible (or, at most, have sensitivity of a very rudimentary kind). The natural "needs to know" govern not only the presence of sensors and sense organs, but what form they assume. There is the need to locate and identify objects of desire, such as food, water, or a mate. There is the converse need to avoid dangers and preserve oneself from enemies. There is the closely related need to navigate and orient oneself. Internal sensations answer to the need to maintain the system (e.g., pain means that damage has been done to the organism). There is also the need to communicate with others, so that needs can be addressed in common.

Similar purposes drive the development of artificial sensors. Radar is used for the location of aircraft and the detection of severe thunderstorms. Various sensors may be used on the battlefield to target or avoid enemy forces. Acoustic and electrical sensors assist in the mapping out of oil and gas fields. Thus, artificial sensors generally answer to practical human needs, directly or indirectly.

This is not always the case, as there is another kind of "need to know", one possessed only by human beings. There is the pure need to know, apart from any practical (i.e., natural or material) consideration. This is the driving force behind the development of sensors in scientific research. If we observe distant galaxies through sophisticated telescopes, this need to know cannot possibly be reduced to mere practical ends. Likewise is the case of aesthetics. For example, what is the purpose of the photometer in a camera? It is to adjust the exposure time so that pictures come out clearly. But this end is subordinated to the essentially aesthetic end of photography -- the desire to have beautiful and pleasing pictures. Although practical in its immediate end, the photometer is pure in its ultimate one. We use our sense organs for both practical and non-practical purposes. The eye that helps me walk through the house without running into walls and locates food in the refrigerator is the same eye that casts its gaze upon a painting or the pages of a book. (How much more pleasant life would be if our sense organs, instead of being primarily adapted to natural ends, were coordinated instead to aesthetic, moral, or scientific ends. Of our senses, hearing, with its great utility in communication, comes closest to this.)

11.3 APPLICATIONS

Let us examine now, very briefly, some potential applications of the generalized notion of sensibility. To repeat, the sensors and the objects sensed are connected to one another and form a system. Thus, the first thing one must define is the structure and means of these connections.

The structure of the interconnections defines a kind of space. The maximum possible "interaction space" for N objects has N-1 Euclidean dimensions. If the relations of the objects are all the same, they will form an N-simplex. This would be the case, for example, if every station on a network is directly connected to every other station. (The 2-simplex is a line segment, the 3-simplex is an equilateral triangle, the 4-simplex is a tetrahedron, and so on.) Spatial relations as we normally consider them, may be considered as a "relation with a magnitude". Thus, the maximum actual interaction space of N objects will be a distortion of the N-simplex contained within a N-1 dimension Euclidean space. Now, the real interaction space will nearly always fall short of this maximum: despite the vast number objects in the universe, for example, there are only three spatial dimensions (four space-time dimensions). Smaller interaction spaces are always a subset of the maximum, and can be obtained by a reduction or "pruning" of the maximum space. This may yield a symmetric and continuous series of possible magnitudes of relations, as in physical space. Or, it may yield discrete and discontinuous values, as on a grid or the branches of a network. Further exploration of this belongs to mathematics and not to sensor theory. Here it is enough to note again that all sensory relations have a spatial aspect.

The connections themselves are not generic. Sensation of its essence forces the relation of objects and sensors into specific channels. There are specific modes of interaction. These modes define the content of the world which the sensor sees. They define what kind of properties of the object are capable of being sensed and the sensory appearance of objects. They define the degree to which the knower can gather knowledge about the known.

We see once again how closely sensor theory verges on systems theory. For the above conditions are also the most important things in the characterization of a system: the structure of possible relations and the nature of the relations themselves. Sensation is the gathering of information about other elements of the system.

The communication of stations on a computer network is an example of this, and an especially good example within computer networks is the growing field of distributed simulation. In distributed simulation, there are a large number of computers connected by a network. Each computer simulates a specific systemic element (an aircraft or the earth's surface, for example) and its dynamic reactions to the other elements. The various stations send out appropriate encoded information regarding their current states, and each station will collect information about the others. The layout of the network is thus the maximum interaction space, and the various kinds of signals define the kinds of interaction. A given station may ignore nearly all signals in order to focus on those few that are of true interest to it. All the basic aspects of sensation are present in a distributed simulation. Information systems gather information about objects outside themselves and act on those object. Communications are restricted to a few possible channels, and each node on the simulation network may have a different kinds of messages (i.e., stimuli) it will detect -- that is, each information system will have its sensory modes. These sensory modes ultimately depend on the purposes of each information system, its "need to know". And the "world" as presented to each information system on the simulation network will have aspects of both space and time.

As mentioned earlier, the communication of organs in the body is another example to which sensor theory can be applied. The geometry of interaction is defined by the circulatory and nervous systems, and may be nearly symplectic (i.e., the theoretical maximum). The modes of interaction are defined by the various chemical substances in the blood and impulses in nerves. We see once more how close certain aspects of sensor theory are to systems theory.

Sensor theory can have applications to the study of society, and this application is by no means limited to human society. A beehive, for example, uses its worker bees as sense organs for the entire hive, to search out sources of nectar. This information is translated into a symbolic system for use within the hive. Similar applications of sensor theory could be made to the study of ecosystems.

Sensor theory may also have importance for economics. In recent years, there has been an increasing awareness that a major function of market relations is the conveyance of information. Prices are indicators of the state of the system and form the basis of future decisions. The flows of goods, services, and money define a space or network. The kinds of flows define the modes of interaction. The monitoring of such flows (for the economy is a dynamic process and not a static thing) may be considered a kind of sensory process.

Many other possible applications of a generalized sensor theory will likely arise in the years ahead with the growth of information and communications technology. Any time that there is a relation of an information system and an object through an intermediary, the label sensation may be applicable.

APPENDIX A

GLOSSARY OF TERMS

(Words in *italics* refer to other entries in the glossary.)

active object an object that actively emits energy, rather than simply reflecting, absorbing, or transforming it.

active protosensor a matched pair of a *protosensor* and a *protoactuator*.

active sensor a sensor which has a means of acting on or illuminating its objects, and thus need not rely on ambient conditions alone. Opposite of *passive sensor*.

actuator a device which converts information into physical energy -- the opposite of a sensor. Actuators, like sensors, are divided into simple, complex, multimode, etc.

additive synthesis an integration of multisensor data by summing the sensor outputs into a single number.

active Argus sensor an *Argus sensor* with active sensing in every mode.

Argus sensor	the largest and most complex sensor theoretically possible, capable of sensing in all modes with infinite extension and resolution. The Argus sensor provides a comprehensive view of the sensory world.
attention	the selective property of *perception*.
attenuation	the decrease of the stimulus intensity due to the medium intervening between the object and the sensor. *Obscuration* is total attenuation.
band	the extent and type of state parameters detected by a sensor: frequencies of light, for example.
binary synthesis	same as monovalent synthesis.
black	a neutral (zero saturation) simple quality, corresponding to low or zero intensity.
carrier	a physical vehicle whose function is to bear an informative pattern. By definition, the nature of the carrier itself is not the object of interest.
complex sensor	a sensor with more than one mode or increment of band, field, or range. Such sensors can always be represented as a combination of simple sensors.
cosmoscience	everything that can be known about the physical world, i.e., the cosmos. An *Argus sensor* could theoretically provide means for cosmoscience.
cropping	a variety of *projection* leading to reduction of the image in extent.

Glossary of Terms

cumulative synthesis   an integration of multisensor data such that only the highest or lowest value (i.e., a magnitude) results.

data   a collection of facts, represented by symbols or numbers, regardless of whether they are of interest or not. (In science, data has the implication of being something quantitative; that is not the case here.)

deflection   dislocation or diversion of the stimulus by the medium. Refraction, reflection, scattering, and diffraction are all examples of deflection.

detector   synonymous with *simple sensor*. The term detector is generally used for isolated simple sensors.

direct fusion   sensor data fusion on the basis of the identities of objects already known by the sensor.

distal stimulus   the *stimulus* (physical interaction) as it is at the object; i.e., at a distance from the sensor.

distortion   a *projection* that modifies the spatial, temporal, intensity, or band increment scaling. Distortion is unavoidable when projecting a higher dimensional object into a lower dimensional image; map projections are the best example.

edge enhancement   a *single image operation* which is intended to raise or highlight sharp changes in intensity (edges), and correspondingly suppress other patterns.

energy measure of the magnitude of a physical action or interaction.

error for sensor theory, error is that which reduces the fidelity of the image to the object, either by adding unwanted information or removing desired information.

extension the size or area of space, time, band, or intensity range covered by a sensor.

exterior illusion an *illusion* due to factors outside the sensor and information system: geometry, the medium through which the stimulus passes, etc.

extrinsic fusion fusion of sensor data on the basis of an extrinsic commonality, such as shared location in space, time, or correlated motion.

field the region of space in which the sensor responds to stimuli. Synonymous with *field of view*.

filter same as *single image operation*.

frame a sensory image or data representation at a single time. Implies membership in a temporal *series* of successive frames, which are excised from a continuous data stream.

frame rate the rate at which discrete data representations or images are sampled from the stimulus data stream.

gray a neutral (zero saturation) simple quality,

corresponding to a moderate intensity evenly distributed across the sensor band parameters. The distinction between gray and *white* is relative and may depend on context.

illusion something appearing to be what it is not.

image the data representation produced by a complex sensor. The term "image" is taken to apply to any sensory mode, not just vision.

image analysis the process of separation of desired patterns or sub-images (i.e., of specific objects) from a larger image. Synonymous with *segmentation*.

image cluster two or more very similar images, which are dissimilar from other images. Clusters, when averaged, yield reference patterns which are signified by a class name or concept.

information data of actual interest: that from which knowledge about something is possible.

information system a system which receives information and makes decisions accordingly.
Computers and nervous systems are both information systems for sensor theory.

input full scale synonymous with *range*.

instrument a device which yields a quantitative measurement. Most, though not all, instruments are sensors.

intelligent object   an active object that can emit intelligent patterns; these are not comprehensible on the basis of the physical properties of the object.

intension   used in two senses. In the first, it is synonymous with the *resolution* of a complex sensor. In the second, it denotes the degree of detail or number of attributes subsumed by a concept.

intensity   the magnitude or quantity of the sensation. Intensity measures the degree of interaction of the sensor and the object.

intentionality   existing in the information system only with reference to the thing known. See §9.5 for further explanation.

interior illusion   an *illusion* due to something inside the sensor or information system: hallucinations, etc.

intrinsic fusion   sensor data fusion on the basis of complex qualities in the objects viewed.

macrocomplex   a *complex quality* (spatio-temporal pattern) for which the extension in quality space and time are consciously perceived.

masking   partial *obscuration*, often with some pattern.

mode   the specific kind of stimulus sensed, such light (vision), sound (hearing), etc. Also pertains to the ability to sense such stimuli.

model	synonymous with *template*.
microcomplex	a *complex quality* (spatio-temporal pattern) for which the extension in quality space or time is not consciously perceived. Typically a repetitive pattern or texture.
monovalent	the combination of several sensor outputs into a single bit of synthesis information. This has the effect of combining several sensor elements to become one larger simple sensor.
multielement sensor	a sensor composed of two or more simple sensors. There are two basic kinds of multielement sensors: *complex sensors* and *multimode sensors*.
multimode sensor	a sensor combining elements or whole sensors belonging to different sensory modes.
multivalent synthesis	a direct combination of sensor outputs into a single representation. Multivalent synthesis yields the most information of all possible data syntheses.
natural object	an *active object* that emits energy purely due to natural processes, e.g., atomic transitions, black-body radiation, etc. See also *intelligent object*.
noise	additive random *error*.
normalization	a *single image operation* which removes extraneous information or detail, simplifying the image. Smoothing is an example.

obscuration	the blocking of the transport of the distal to the proximal stimulus by some opaque intervening object.
parameters	the term is used in two ways in this book: 1. the state descriptors of a physical interaction, such as frequency and polarization of light. 2. the properties of a sensor, such as band, field, and range.
passive object	an object that merely reflects, absorbs, or transforms energy. Opposite of an *active object*.
passive sensor	a sensor which merely receives energy from its ambient environment. Opposite of *active sensor*.
pattern	for sensor theory, a pattern is an *image* or portion of an image: a spatial or temporal array of intensities and qualities.
pattern matching	comparison of two images to yield a scalar measure of their similarity.
pattern recognition	identification or classification of a pattern by comparing it with one or more reference patterns.
perception	the knowing awareness of the subject, as distinguished from *sensation*.
perspective	the partial viewpoint of a finite sensor due to *projection*.
pixel	a pointlike or discrete element of an *image*. Images are built from combinations of pixels.

primary qualities	in epistemology, those qualities which supposedly belong to the object itself: shape, extent, motion, etc. Similar to *macrocomplex qualities*.
projection	the transport of the distal to the proximal stimulus and its limitation to the sensor conditions -- e.g., as in the projection of a shadow on a wall.
protoactuator	an idealized *actuator* of infinitesimal bandwidth, field of action, and intensity range. Opposite of *protosensor*.
proto-object	an infinitesimal (or discrete) point into which a *sensory object* can be virtually or actually decomposed. A proto-object is to the sensory object what a pixel is to an image.
protosensor	an idealized *simple sensor* of infinitesimal bandwidth, field of view, and intensity range, with binary output.
proximal stimulus	the *stimulus* (physical interaction) as received at the sensor.
quality	specifies the kind of intensity as perceived. The term quality is always used in this book to mean sense quality: an sensible attribute of an object. Qualities can correspond to the parameters of the carrier (*simple qualities*) or can be spatio-temporal patterns of intensities *complex qualities*).
range	the extent of intensities to which the sensor will respond, from a minimum *threshold* value to a

maximum *saturation* value.

rebinning a *projection* which reduces the resolution of the image with respect to the sensory object, having the effect of blurring or smearing out the object.

receptor generally synonymous with *simple sensor*. The term receptor is normally used for the elements of a biological sense organ.

reference pattern a pattern used as the archetype or "best example" in pattern matching; other patterns are compared to it for classification. Reference patterns are likely to be *templates*. A class name or concept will correspond to the reference pattern.

representation a pattern, symbol, or number within an information system.

resolution the detail or number of increments possessed by a complex sensor in space, time, band, or range.

saturation used in two senses: (1) an intensity value above which either the sensor cannot respond or the information is meaningless: a maximum intensity value, and (2) the purity of a simple quality.

secondary mode the ability of a sensor to detect stimuli not belonging to its intended or primary mode. Some examples of secondary modes are feeling the warmth of sunlight on one's skin, seeing colors from pressing on the eyeballs, and feeling the low frequency rumble of a thunderclap.

secondary quality in epistemology, a quality which does not belong to

	the object, but to the mind of the observer, such as colors, sounds, smells, etc. Equivalent to *simple qualities*.
segmentation	same as *image analysis*.
sensation	the process of conversion of a physical action of an object into a representation within an information system: the actualization of a sensory representation in relation to an object.
sense organ	a *sensor* possessed by a living being.
sensibility	the possibility of sensation
sensor	a device that transforms energy into information. All sensors are *transducers* (q.v.) in that they convert one kind of energy into another. However, the output of a sensor is symbolic of the kind of *stimulus* (q.v.). A sensor makes sensation possible for an information system.
sensor element	generally synonymous with *simple sensor*. The term sensor element is used for simple sensors that are the parts of a larger *complex sensor*.
sensor fusion	the problem of joining several sensors or sensor elements together into a single sensor system.
sensor suite	taken as synonymous with *multimode sensor*. Implies the combination of several separate sensors into one system.

sensory object	the *distal stimulus*, for all modes, coming from an object; the bundle of physical interactions corresponding to the object -- hence, its sensory appearance.
series	a set (manifold) of objectively related and temporally successive data *frames*.
shadowing	*obscuration* of the source of illumination.
simple sensor	a sensor with a single undifferentiated mode, band, field of view, and intensity range.
signal	a stimulus that is a communication or came from an intelligent source.
signature	the unique sensory appearance of a given physical substance or object.
single image operation	a systematic modification or processing of an image that requires one image as input and produces one image as output. Synonymous with *filter*.
span	synonymous with *range*.
spatio-temporal fusion	same as *extrinsic fusion*.
static sensor	a sensor whose parameters -- band, field, range -- are fixed.
stimulus	a physical influence that can be sensed -- light, sound, etc. They are divided into *distal* and *proximal*

stimuli.

template an image formed from averaging two or more other images, especially when these images are very similar. Equivalent to *model* and used for *reference patterns*.

threshold the minimum intensity value to which a sensor will respond.

thresholding a *single image operation* that suppresses all parts of an image below a specified intensity. Multiple thresholding yields intensity contours.

time lag projection due to the fact that the transport of the stimulus from the object to the sensor cannot occur instantaneously. Thus, the object is viewed as it was at the moment the stimulus (physical interaction) departed.

transducer a device that converts one kind of energy into another: light into electricity, etc. All sensors are transducers, but not the reverse.

variable sensor a sensor whose parameters can be changed depending on conditions -- the opposite of a static sensor.

white a neutral (zero saturation) simple quality, corresponding to a high intensity evenly distributed across the sensor band parameters: white-light combines all colors, white noise combines all pitches, etc.

world image the totality of one's stored images in relation to one another.

NOTES AND REFERENCES

CHAPTER ONE

[1] I am *not* asserting that the human mind and its cognitive processes are reducible to a node of sensory information. A discussion of the mind-brain problem is well beyond the scope of this book. Sensor theory is only concerned with the origination, validity, and representation of sensory information, and for its purposes, the mind is an information system.

Likewise, sensor theory can neither prove nor disprove parapsychological claims of extrasensory perception (e.g., telepathy). It merely can show that such modes of cognition, if they exist, are not sensation, because they do not force perception into specific channels.

[2] Jacob Fraden, *AIP Handbook of Modern Sensors* (New York: American Institute of Physics, 1993), p. 1,2.

[3] John Brignell and Neil White, *Intelligent Sensor Systems* (Philadelphia: Institute of Physics Publishers, 1994), p. 33.

[4] The assumptions are just that: provisional. Scientists and engineers will have no problem with such physical assumptions. To address some philosophical concerns, however, Chapter 11 shows how the notion of sensors and sensation can be extended to non-physical situations. General sensor theory does not depend on the structure of the physical world.

[5] William N. Dember and Joel S. Warm, *Psychology of Perception* (New York: Holt, Rinehart and Winston, 1979), p. 3.

[6] Ibid., pp. 6-8.

[7] Ibid., p. 15.

[8] Active sensors are sometimes defined as a sensor that has an external power source, and thus need not rely upon the energy contained in the stimulus. (See Fraden, p. 3.) Such a sensor would be passive by my definition.

[9] See, for example, the definitions in Heinrich Niemann, *Pattern Analysis and Understanding* (Berlin: Springer-Verlag, 1990), p. 206.

[10] Robert F. Schmidt, editor, *Fundamentals of Sensory Physiology* (Berlin: Springer-Verlag, 1986), pp. 2-4.

[11] Ibid., pp. 68-71.

[12] Ibid., p. 115.

[13] Ibid., p. 250.

[14] Ibid., p. 31.

[15] L. Bolis, R. D. Keynes, S. H. P. Maddrell, editors, *Comparative Physiology of Sensory Systems* (Cambridge: Cambridge University Press, 1984), p. 405ff.

[16] Ibid., p. 301ff.

[17] Ibid., p. 115ff.

[18] Ibid., p. 3ff.

[19] Ibid., p. 455ff.

[20] Ibid., p. 497ff.

[21] See, for example, Fraden, *op cit.*

CHAPTER TWO

[1] Fraden, *AIP Handbook*, p. 18.

[2] Schmidt, *Fundamentals of Sensory Physiology*, pp. 157, 177.

[3] An example of this are pressure forces or temperature. Although they are measured at a point, they refer to an entire system. On a grander scale, there is the cosmic infrared background radiation found everywhere in the universe.

CHAPTER THREE

[1] Tom N. Cornsweet, *Visual Perception* (New York: Academic Press, 1970), p. 129.

[2] R. Schmidt, ed., *Fundamentals of Sensory Physiology*, p. 156.

[3] Faber Birren, *Principles of Color* (West Chester: Schiffer, 1987), p. 50.

[4] *Sensory Physiology*, p. 73.
[5] Dember and Warm, *Psychology of Perception*, p. 93.
[6] Charles Blilie, *Visible/IR Comprehensive Sensor Simulation (VICSSN) Technical Reference*, VI-2091 (July 1993), §4.
[7] Aristotle, *Categories*, 8 (9a).
[8] Aristotle, *On the Senses*, 4 (445b).
[9] Committee on Colorimetry, *The Science of Color* (Washington, DC: Optical Society of America, 1973), p. 67.
[10] *Sensory Physiology*, pp. 180-181, 219.
[11] Dember and Warm, p. 429.
[12] C.f., e.g., Russell L. DeValois and Karen K. DeValois, *Spatial Vision* (Oxford University Press, 1990).
[13] Marr, *Vision*, p. 43.
[14] *Sensory Physiology*, pp. 22-25.
[15] C.f., e.g., Eugene Hecht and Alfred Zajac, *Optics* (Menlo Park: Addison-Wesley, 1979), p. 467ff.
[16] H.J.A. Dartnell et al., in J.D. Mallon, editor, *Colour Vision* (New York: Academic Press, 1983).

CHAPTER FOUR

[1] David L. Hall, *Mathematical Techniques of Multisensor Data Fusion* (Boston: Artech House, 1992), p. 2.
[2] Ibid., pp. 22 - 27.
[3] H.F. Durant-Whyte, "Sensor Fusion: when more means better", in K.T.V. Grattan, ed., *Sensors: Technology, Systems and Applications* (New York: Hilger, 1991), p. 411.
[4] Hall, pp. 3-5.
[5] Ibid., p. 36.
[6] Durant-Whyte, p. 411.
[7] Aristotle, *On the Soul*, II.6 (418a).
[8] Durant-Whyte, p. 412.
[9] Hall, p. 24.
[10] On binocular vision, c.f., e.g., Marr, pp. 112-119. It should be noted

that binocular viewing is not the only way of recovering depth information. One can use, for example, sonar, radar, and laser rangefinders; there are also the techniques of shining patterned light on objects to delineate their shapes and the focussing/defocussing of an image by varying the focal length of a camera (Niemann, p. 122). But all of these are *active* sensing techniques: to a purely passive sensor, binocularity is the only option.

[11] Barry E. Stein and M. Alex Meredith, *Merging of the Senses* (Cambridge: MIT Press, 1993), p. 1.
[12] Ibid., pp. 4, 7.
[13] Ibid., p. 2.; Dember and Warm, p. 10.
[14] Stein, p. 3.

CHAPTER FIVE

[1] Fraden, *AIP Handbook*, p. 28ff.
[2] C.f., e.g., Brignell and White, *Intelligent Sensor Systems*.

CHAPTER SIX

[1] Immanuel Kant, *Critique of Pure Reason*, tr. by N. K. Smith (New York: St. Martin's Press, 1929) Bxii - Bxiv. Yet even Kant does not light upon the problem of active sensation.

CHAPTER EIGHT

[1] Moltke S. Gram, *Direct Realism: A Study of Perception* (The Hague: Martinus Nijhoff, 1983), pp. 2, 52, 157.
[2] Some of these examples are borrowed from J.L. Austin, *Sense and Sensibilia* (London: Oxford University Press, 1962), pp. 20-21, 42 in his discussion of common illusions used as objections to veridical sense-perception.
[3] For a discussion of the sampling theorem with regard to images, see Anthony Vanderlugt, *Optical Signal Processing* (New York: Wiley Interscience, 1992), pp. 3 - 10.
[4] Fraden, *AIP Handbook*, pp. 21-25.
[5] ibid., p. 22.
[6] See, e.g., Richard E. Blahut, *Principles and Practice of Information*

Theory (Reading: Addison-Wesley, 1987).

CHAPTER NINE

[1] Aristotle, *On the Soul*, II.7 (419a); *On the Senses*, 3 (439a).

[2] Aristotle, *On the Soul*, II.12 (424a).

[3] For a discussion of the Aristotelian sensible species in a Thomistic context, see George Klubertanz, *The Philosophy of Human Nature* (New York: Appleton-Century-Crofts, 1953); Louis Regis, *Epistemology* (New York: Macmillan, 1959).

[4] Aristotle, *On the Senses*, 4 (445b).

[5] Aristotle, *On the Soul*, III.1.

[6] The philosopher Husserl held that the "demotion" of secondary qualities by Galileo was one of the major turning points in Western thought. Edmund Husserl, *The Crisis of European Sciences and Transcendental Phenomenology* (Evanston: Northwestern University Press, 1970), pp. 23, 34, 37.

[7] Harmon Chapman, *Sensations and Phenomenology* (Bloomington: Indiana University Press, 1966), pp. 45-46.

[8] H.H. Price, *Perception* (London: Methuen, 1964), p. 3. (as quoted in Gram)

[9] Austin, *Sense and Sensibilia*, p. 61; Gram, *Direct Realism*, pp. 174-181.

[10] Chapman, p. 3f.

[11] Thomas Reid (K. Lehrer and R. Beanblossom, eds.), *Inquiry and Essays* (Indianapolis: Bobbs-Merrill, 1975); James McCosh, *The Scottish Philosophy* (Hildesheim: Georg Olms, 1966), pp. 158-224.

[12] Gram, *op cit.*

[13] Robert Boynton, "Human Color Perception", in K. N. Leibovic, editor, *The Science of Vision* (New York: Springer-Verlag, 1990), pp. 222-223.

[14] I am discounting here phenomena within the information system itself or noise in the sense organs (e.g., ringing in the ears). Section 9.5.2 deals with the question of interior illusions or hallucinations.

[15] The properties of the object are, however, exhausted by the qualities available to the Argus sensor.

[16] The Cartesian "evil genius" is in many ways a virtual reality problem. It might not be difficult to fool the unaided human senses -- to make us perceive a false reality and thus draw false inferences from it. But even such an evil genius could fool an Argus sensor only by actually creating a new reality in every detail. At that point, virtual reality passes over into reality.

[17] Klubertanz, pp. 68-69.

[18] Harmon, p. 99ff.

CHAPTER TEN

[1] For details of edge enhancement in human vision, see Marr, *Vision*, p. 54ff; Schmidt, *Sensory Physiology*, pp. 22-23 and pp. 176-178; Dember and Warm, *Psychology of Perception*, p. 220.

[2] Dember and Warm, p. 251.

[3] Niemann, p. 34.

[4] Schmidt, p. 115.

[5] Niemann, p. 75.

[6] Marr, p. 47.

[7] Dember and Warm, p. 254ff.

[8] Niemann, p. 75.

[9] Marr, pp. 37, 41-42, 52-53.

[10] Schmidt, p. 188.

[13] John Aliomonos and David Shulman, *Integration of Visual Modules: An Extension of the Marr Paradigm* (San Diego: Academic Press, 1989).

[12] Marr, p. 270f.

[13] Dember and Warm, p. 124.

[14] Ibid., pp. 127-128.

[15] I am well aware that time is unified with space in the physical theory of relativity; whether that fact has any relevance to the association of images is questionable.

[16] Marr, p. 32.

[17] Dember and Warm, p. 227.

[18] Marvin Minsky and Seymour Papert, *Perceptrons: An Introduction to Computational Geometry* (Cambridge: MIT Press, 1988).

[19]   Niemann, p. 187.
[20]   Ibid., 9ff.
[21]   Dember and Warm, p. 269.
[22]   Marr, p. 318.
[23]   Dember and Warm, p. 11.
[24]   For a discussion of the applications of fuzzy logic to pattern matching, see James C. Bezdek, Sankar K. Pal, *Fuzzy Models for Pattern Recognition* (New York: IEEE Press, 1992), pp. 1-25.
[25]   Austin, *Sense and Sensibility*, p. 121.
[26]   D. Dutton, et al., *Spectra-Physics Laser Technical Bulletin* 3, as reproduced in Hecht and Zajac, p. 469.

INDEX

A

active object, 105, 199, 203, 205, 206
active protosensor, 96, 106, 199
active sensor, 10, 91, 92, 93, 94, 95, 96, 98, 105, 107, 131, 146, 163, 199, 206
actuator, 11, 93, 94, 95, 96, 153, 163, 192, 199, 207
additive noise, 128
additive synthesis, 45, 55, 199
America, 215
Argus Sensor, vi, 79, 98
Aristotelian theory of sense qualities, vii, 137
artificial intelligence, xii, 78, 99, 167, 176
artificial sensors, xi, xii, xiii, xiv, 5, 10, 24, 25, 26, 27, 41, 69, 70, 75, 76, 83, 88, 122, 125, 127, 157, 175, 195
attenuation, 117, 123, 124, 131, 200

B

binary synthesis, 45, 200
Biological Sensory Integration, vi, 76

C

cerenkov detector, 26
cloud chamber, 26
cochlea, 20

complex quality, 56, 65, 72, 204, 205
complex sensors, extension and intension of, vi, 66
corruption, 46, 132
cosmoscience, 99, 107, 200
cropping, 117, 118, 200

D

Descartes, 35, 140, 141, 143, 157
direct fusion, 72, 201
distal stimulus, 6, 103, 109, 113, 132, 201, 209
distortion, 117, 118, 120, 127, 196, 201
Dynamic Aspects of Sensation, vi, 81
Dynamic Response of Sensors, vi, 82

E

edge enhancement, 169, 170, 201, 218
electric fields, 13, 24, 25
exterior illusion, 119, 120, 121, 145, 158, 202
extrinsic fusion, 72, 202, 210

F

Forward-Looking Infrared (FLIR), 86
frame rate, 84, 85, 88, 89, 133, 158, 202
fusion, spatio-temporal, 71, 72, 73, 74, 75, 210

G

Galileo, 25, 140, 217
gamma rays, 15, 23, 26, 79
geiger counters, 26
general sensor theory, xii, xiv, xv, 79, 108, 130, 143, 193

H

hallucinations, 158

I

image analysis, xv, 169, 177, 178, 179, 203, 209
image cluster, 186, 189, 203
image segmentation, vii, 175
imaging infrared sensors, xi
Indiana, 217
information system, 3, 5, 11, 28, 35, 36, 46, 58, 64, 74, 82, 83, 84, 88, 89, 94, 96, 99, 111, 115, 132, 134, 137, 144, 146, 147, 150, 151, 153, 156, 157, 158, 162, 163, 167, 168, 170, 173, 175, 182, 193, 197, 198, 202, 203, 204, 209, 213, 217
infrared detector response function, ix, 54
infrared light, 15, 22, 26, 124
input full scale, 30, 203
intelligent object, 132, 203, 205
intensity levelsm synthesis of, 47
intentionality, 137, 139, 160, 161, 162, 163, 204
intrinsic fusion, 72, 74, 75, 204

K

Kalman filter, 73

L

lateral line, 22
Locke, 107, 141, 143

M

macrocomplex, 53, 64, 65, 73, 74, 75, 144, 147, 149, 150, 162, 204, 206
magnetic fields, 13, 25
Marr, David, 177
masking, 123, 204
mass spectrometers, 26
mechanoreception, 19, 21
mechanoreceptors, 21, 30, 41
mercury, 25
microcomplex, 53, 63, 64, 74, 75, 144, 147, 149, 150, 151, 162, 177, 205
multimode sensor image, vi, 78
multivalent synthesis, 45, 205

N

natural science, 96, 104, 110, 111, 155
normalization, 173

O

object, passive, 105, 206

P

pattern matching, 175, 178, 185, 206, 208, 219
pattern recognition, 65, 71, 74, 78, 167, 168, 178, 179, 182, 183, 184, 185, 186, 206
pheromones, 22
photodetector, 10, 37, 38, 39
phototropism, 5
pit organ, 22
pixel, 51, 52, 54, 86, 128, 134, 135, 146,

206, 207
polarization states, 61, 62
positional fusion, 72, 74, 75
primary qualities, 56, 64, 65, 73, 139, 141, 142, 143, 144, 206
primary tones, 20
protoactuator, 94, 96, 106, 199, 207
proto-object, 104, 105, 106, 207
protosensor, xiii, xv, 10, 29, 30, 31, 32, 33, 34, 35, 36, 37, 38, 41, 43, 56, 66, 79, 80, 82, 96, 106, 108, 137, 147, 162, 199, 207
proximal stimulus, 6, 103, 113, 114, 116, 205, 207

Q

quasi-symplectic space, 181

R

rebinning, 117, 118, 120, 122, 148, 207
receiver/decoder, 132
reference pattern, 179, 182, 183, 184, 185, 186, 188, 203, 206, 208, 210

S

sampling theorem, 89, 122, 216
secondary mode, 125, 129, 208
secondary quality, 208
segmentation, 168, 175, 176, 178, 179, 180, 186, 188, 203, 209
sense organ, xii, xiv, 5, 10, 17, 19, 21, 22, 24, 25, 27, 41, 42, 51, 56, 69, 84, 86, 88, 124, 125, 127, 137, 138, 140, 148, 149, 157, 165, 168, 192, 194, 195, 197, 208, 209, 217
sensor fusion, 42, 69, 73, 74, 75, 76, 78, 209
sensor, multielement, 10, 41, 42, 47, 205
sensor, multimode, xiii, xv, 10, 42, 69, 71, 76, 78, 79, 84, 144, 181, 205, 209
sensor, passive, 10, 79, 92, 95, 199, 206, 216
sensor, simple, 9, 11, 29, 31, 37, 38, 39, 41, 42, 45, 47, 56, 66, 80, 200, 201, 205, 207, 208, 209, 210
sensor, static, 11, 210, 211
sensory object, xv, 72, 103, 104, 105, 106, 107, 108, 109, 110, 111, 113, 115, 116, 117, 118, 119, 120, 121, 123, 124, 131, 132, 137, 145, 148, 154, 162, 163, 207, 209

shadowing, 210
silkworm moth, 22
simple quality, 56, 62, 71, 153, 200, 202, 208, 211
single image, 43, 84, 86, 169, 170, 173, 201, 202, 205, 210, 211
spark chamber, 26
Spatio-temporal (extrinsic) Fusion, vi, 72
Stevens power law, 52
synaesthesia, 77, 130, 158

T

telecommunications, 130
thermoreception, 19
thermostat, xi, 38
time lag, 117, 118, 147, 211
Time lag, 118
transducer, 5, 211

U

ultrasound, 24, 26
ultraviolet light, 15, 26, 117
United States, 117

V

variable sensor, 11, 87, 88, 94, 211
voltmeters, 26

W

Weber-Fechner law, 52
world image, 182, 186, 188, 189, 211